F. E. Potts'
Guide To
BUSH FLYING

Concepts and Techniques for the Pro

ACS Publishing

Tucson, Arizona

F. E. Potts' Guide To BUSH FLYING:
Concepts and Techniques for the Pro

First edition

ACS Publishing
3906 West Ina Road, Suite 200-211
Tucson, AZ 85741

E-MAIL
Internet:
ACS Publishing: acs@fepco.com
F. E. Potts: fepotts@fepco.com

UUCP:
F. E. Potts: uunet!fepco!fepotts

CompuServe:
F. E. Potts: 71570,276

Cataloging in Publications Data

(Prepared by Quality Books, Inc.)

Potts, Fred E.
F. E. Potts' guide to bush flying:concepts and techniques for the pro—1st ed.
p. cm.
Includes index and glossary
ISBN 0-9635210-1-2
Library of Congress 92-075524
1. Bush flying. I. Title. II. Title: Guide to bush flying.
TL711.L7P68 1993 629.132'5216
QB193-20124

Book production by Comp-Type, Inc., Fort Bragg, CA 95437
Manufactured in the U.S.A.

Dedication

This book is dedicated to the many fine customers I have had over the years, and to all the pilots who fly in remote areas, wherever those areas may be.

Acknowledgments

I wish to express my appreciation to the three fine women who accompanied me during different stages of this journey: Dolly, who was first, and the most adventurous; Belphi, the fiery warrior who came second; and Elaine, the contemplative one who is always seeking answers to the deepest questions. Without their help and encouragement, this book would never have been written.

I also wish to thank Mike Busch, Denny Cunningham, John Deakin, Barry Schiff, Dick Zeitlin, and Margalit Zur for reading the manuscript in its next-to-last draft, and providing me with their valuable comments and suggestions.

I also wish to express my appreciation to Robert Dubner, who provided me with the official names of the tiedown knots I have been using all my life (and provided a rationale for using the descriptive "Hurricane Hitch" to describe the series of knots used). He also provided the approved nautical terminology, for describing how to tie these knots, that I used in Chapter 27.

Finally, I wish to express my appreciation to John Fremont and Cole Loftus, both of whom struggled mightily in the realm of copy editing to get my unruly text into shape for publication.

Warning & Disclaimer

This book contains methods and/or flying techniques that have worked successfully for the author (as well as for other professional bush pilots) for many years. However, for those *without* the required skill, experience, or judgment, these techniques *(because they utilize the edges of the airplane's performance envelope, close to the stall; or require operation low to the ground; or push weather)* are very dangerous. They cannot be properly or safely learned just by reading a book, but require practical, hands-on instruction. Therefore, pilots wishing to learn these advanced techniques should take thorough instruction from a *Certified* Flight Instructor *qualified in bush techniques,* then understand [see FAR Part 91.3(a)] that pilots who elect to use these techniques must of necessity accept all responsibility for their own and their passengers' safety and property.

Every effort has been made to make this book as complete and accurate as possible. However, there may be errors, both typographical and in content. Therefore, this book should only be used as a general guide as to how some professional bush pilots, including the author, operate, and not as an instruction manual.

The purpose of this book is to educate and entertain. The author and ACS Publishing shall have neither liability nor responsibility to any person or entity with respect to loss or damage caused, or alleged to be caused, directly or indirectly, by the information contained in this book.

Credits

Dan B. McCarthy's *Foreword,* in slightly different form, was first published in Volume 3, Number 18 of *Alaska Flying* under the title *"By The Numbers" Bush Pilot.*

Various parts of Chapters 12 and 13 were first published in Volume 20, Number 4 of *Plane & Pilot* under the title *Common Sense For the Uncommon Cold.*

Chapter 15 was, in modified form, first published in Volume 7, Number 1 of *IFR Magazine* under the title *RMI Resurrected.*

The Jeppesen Chart used in Chapter 15 courtesy of Jeppesen Sanderson.

Chapter 24 was, in slightly bowdlerized form, first published in Volume 4, Number 26 of *Alaska Flying* under the title *Survival Fit & Kit.*

Chapter 25 was first published in Volume 4, Number 20 of *Alaska Flying* under the title *Optical Illusions.*

Various parts of Chapter 26 were first published in vastly different form in Volume 4, Number 23, and Volume 4, Number 24 of *Alaska Flying* under the title *Contour Flying.*

Table of Contents

Section II: Flying Techniques

Section III: Images

Section IV: Appendices

List of Illustrations

* All photos by author, unless otherwise noted.

Foreword

by Dan B. McCarthy

"By The Numbers" Bush Pilot

Consider his statistics: 22 years of flying in many of Alaska's most demanding regions. Some 17,000 hours at the controls of float, ski, and wheel-equipped single-engine airplanes. No accidents. F. E. Potts flies like a craftsman, respecting his airplanes, his territory, and, above all, common-sense safety measures. To use his own expression, he flies "by the numbers."

"Despite stories one hears," Potts says, "the most effective way to fly an airplane in the bush is scientifically—by the numbers. This allows us to obtain maximum performance, though it does require a somewhat higher level of knowledge and skill than normally encountered."

Potts is no fence-sitter when expressing his feelings about aviation. "My opinion is that flying in the North is the best in the world," he says matter-of-factly. His reasons are well-justified after his lengthy career as a bush pilot, 17 years of which were spent in one of the most beautiful and rugged parts of Alaska—the Wrangell Mountains east of McCarthy.

"I started a small flying service there with an antique Super Cub I managed to purchase with money my wife and I had saved through careful living—mostly off the land."

People who wanted to go into that area for mineral exploration, or to hunt big game, composed his principal market. But civilization finally caught up with him when a road was built into the area and it was turned into a national park, so in 1984 he decided to retire from the commercial flying business (FAR Part 135) and devote himself more to writing. He sold his property to a group of investors headed by his very first customer in Alaska. "He and a friend went out and got themselves a record-book ram. They remained customers for years."

In the spring of 1985 he moved his base of operations to Arizona, where he now has a home and hangar on a private airstrip in the Sonoran Desert. But that doesn't keep him from the North, which he visits regularly, either in his Cessna TU206G or his Beechcraft B36TC, depending on the mission.

Potts and his first wife, Dolly, settled in Alaska in 1962. They were attracted to the state through natural curiosity and impulse. "I was 28 then. I've been a licensed pilot since I was 17, so it was obvious what would be my new trade."

To satisfy this curiosity they traveled about halfway around the world via commercial airlines and touched down in Anchorage. "I have lived and traveled extensively in North Africa, the Middle East, and India," Potts said. "In 1962 Dolly and I happened to be somewhere in backcountry Jordan, traveling among the Bedouin. One afternoon, with the temperature well above 100°, we were sitting under the awning of the chief's tent awaiting the cool of the evening. While passing time and drinking coffee, I was translating the text of a gun catalog from English into Arabic for our companions—a gift I had brought them, but not previously read. Guns were one of their primary interests.

"As I was turning the pages, and answering their questions while we discussed the relative advantages and disadvantages of various guns and calibers, I came to a picture of two hunters in Alaska. They were somewhere on the Kenai Peninsula, posing with their moose in the gently falling snow, rifles in hand, the dim outline of spruce trees barely visible behind them. To one side lay their packboards, waiting, as did a meatsaw resting on top of one of their packs. It was obvious they would be eating well that winter.

"I looked up from that picture of virgin wilderness, then out across the barren desert, through a cloud of blowing sand and dust, flies shimmering in the heat waves. My gaze drifted across the scene until it reached my wife. She nodded, saying nothing.

"So, as though nothing momentous had happened, I finished the translation, allowed our visit to come to its courteous end, and about a week later we were on our way to Anchorage. We arrived in Alaska with the clothes on our backs, my cello—hardly the best tool for earning a living in Alaska—and $18."

They spent their first winter in a tent in the Cooper Landing area. Later they moved to Chitina, and after two years in Alaska they settled in the Wrangell Mountains.

By 1971 he was one of the first 13 pilots in Alaska to have FAR Part 135 single-engine instrument authorization to carry passengers and cargo for hire. He used a Cessna 180 for instrument flying, and also had IFR instrumentation in a Super Cub because of a few special jobs it would permit him to do that could not be done otherwise.

In 1972, Potts was offered a contract to fly for an expedition in East Africa with archaeologist L. S. B. Leakey, who was searching for traces of early man. The expedition was funded by the L. S. B. Leakey Foundation and the Guggenheim Foundation. "They couldn't find anyone in East Africa who could fly the terrain Leakey needed to operate in using fixed-wing equipment. And the cost of helicopter support would take them over budget."

Foundation photographs and maps indicated that, even at the altitude where they would be operating, it would be easy flying in comparison to much of the terrain Potts operated off of in Alaska.

A contract was signed, and Potts and his wife left for New Orleans in their C-180, which was to be shipped from there to Nairobi by sea. Their plan was to accompany it, enjoying a rare vacation in the process. But while they were enroute, Dr. Leakey suffered a fatal heart attack. So the Leakey Foundation instructed them to proceed to their headquarters in Los Angeles.

Without its leader, the expedition was in serious trouble, not only from the loss of knowledge and skill, but from negative publicity as well—always important to funding foundations.

Consultations and attempts to get a substitute leader suitable to the Guggenheim Foundation dragged on for months, and then a year. Eventually the expedition was canceled. Potts philosophically accepted a generous cancellation fee and once again surrendered to his love of the arctic. As he explained it:

"Along with many Northerners, I share the characteristics needed for developing new frontiers: a love of adventure and exploration, a desire to build, to achieve, to cope with difficult problems and come up with pragmatic solutions. For the truth of the matter is, the North is not just a place to live and work, but a state of mind."

Introduction

Alaska has a reputation for being a risky place to fly, a place so dangerous that many insurance companies refuse to provide coverage for flights going into the State.

Insurers cite the roughness of the terrain, the narrow and stormy mountain passes, the vast stretches of desolate tundra, the fast-changing weather systems, the winter cold, the snow and ice, the lack of widespread weather information, the lack of terminal radar except for a couple of big cities and a few military bases, the remote nature of many of the towns and villages, the non-standard airstrips, and the lack of communications.

They also cite as a problem the mindset of many who fly the bush, considering the area one that attracts wild men—men with an oversupply of testosterone and an overly macho psychology. To prove their point, they have accident statistics going back for years, and their case is strong.

There is a fair amount of truth in all this, but it is not the whole truth by any means. It is my considered opinion, based on better than 22 years of flying the remote bush, that, through knowledge and judgment, northern bush flying can, for most pilots, be as safe as flying in the Lower 48. And for some pilots (those with advanced levels of skill and judgment), it can be considerably safer, for there is far less traffic, and therefore far less risk of a midair collision.

Accidents that do occur in the bush are most often the result of pilot error, usually caused by a gross deficiency in judgment.

In 1974, after a friend joined the long and depressing list of those I have known who died in preventable accidents, I decided to write this book. My hope was not only to provide a knowledge base to assist in making judgment calls, but to provide as well a psychological perspective that would be conducive to survival.

This book has taken me a long time to write, for I am primarily a pilot, not a writer, and flying has always been my priority. Along the way, I've had to make numerous decisions, two of which need to be noted here:

My first problem was deciding what to cover in this book. There are good books on the basics of float, instrument, and mountain flying; therefore, I decided to concentrate on subject matter unavailable elsewhere. In this book I touch upon these subjects only in passing, when their inclusion enhances the subject under discussion.

I next had to decide what minimum level of knowledge and skill to assume the reader would bring to this book. The text covers a broad gamut of techniques, ranging from what a pilot with a FAA Commercial Pilot's Certificate could be expected to handle, to skills so advanced they are only suitable for the seasoned pro. So this book takes the Commercial level of skill as its basic starting point, and goes from there.

This book focuses quite a bit on the "pro" attitude, and a definition of pro needs to be made. Pro, as used in this context, *does not* mean a pilot who gets a paycheck for flying under FAR Part 135. Nor does it mean a pilot who has a FAA ATP Certificate. It refers to an attitude towards flying—the thoughtful, careful approach that is so necessary for safe operations.

FlightSafety International claims that the best safety device in any aircraft is a well-trained pilot; I believe the best safety device is a thoughtful, aware, careful pilot who knows how to exercise judgment. Training is very important (and dare not be skimped), but attitude is even more important.

A Note on Style and Structure

He flew past it at 50 knots and half-flaps, checking, and it was as he remembered: 340 feet of narrow, doglegged gravel bar, surrounded by trees and brush in a deep canyon. Along one side the rocky creek from the glacier flowed; once, long ago, during a period of high water, it had created this landing area; eventually, it would destroy it. But for now the gravel bar was still usable, which was fortunate, for it was the only place within miles of where his customer needed to go that an airplane could be set down.

In Alaska, you use what you have and are thankful it is available. For here, a true airport is considered almost a decadent luxury.

"Do you think it's okay?" his passenger called worriedly from the rear seat of the Super Cub. "Looks pretty short to me."

The pilot, looking at the strip, nodded; and, as he did, a picture formed within his mind—the approach path, with the various "windows" along it, necessary to bring him to the threshold at the speed and altitude necessary for a safe landing. "Yeah," he agreed. "It's short—and rough. But good enough. We're going in."

He made his turn, headed downwind, maintaining his 50 knots at an altitude of about 20 feet above the trees; then, on coming abeam the point where he needed to enter the creekbed and descend below the trees, he turned once again, heading back at an angle. As he reached the edge of the forest, he smoothly came back on the power, carefully maintaining airspeed with attitude (elevator), and began his descent.

The creekbed twisted like a snake, its width roughly twice the airplane's wingspan, and the Super Cub followed, going lower and lower, airspeed locked on target, subtle alterations of power controlling descent. Along the *edges, in patches of sand, were the tracks of moose and bear, clearly visible to the passenger, and above them a raven circled. Then they were going around the final bend, this one narrower, and as they did, while in the turn, the pilot applied full flaps and slowed to 1.1 V_{SO}, using his sensitivity of feel for flying on the very edge of the stall, a high level of skill in this area of the envelope being one of the primary tools of the bush pilot's trade.*

In front of them the strip now lay, visible for the first time since they began the approach, the curve of its dogleg obvious, thick brush at its far end. As they neared the threshold and entered ground effect, the pilot raised the nose slightly, while smoothly adding a touch of power, and the Super Cub slowed even more.

Then the main gear was crossing the edge of the shallow bank, hardly a foot above its surface, and, as it did, the airplane's tail still over water, the pilot came back on the power and the Super Cub settled, touching gently, its large, soft, high-flotation tires now rolling smoothly over the rocks. Then the flaps were off—the brakes, as is proper bush technique, having been partially engaged just prior to touchdown—and they were tracking around the dogleg, coming to an easy stop. In front of them, close, but not too close for comfort, the brush waited, a common hazard in the bush. For, as is often the case with off-airport landing areas in Alaska, this was a one-way strip.

There are many factors involved in an off-airport landing besides just the basic flying techniques, and this excerpt from Chapter 17 gives a feel for a typical approach and landing under good terrain and weather conditions.

But, before the reader can safely attempt similar landings, he needs to have a lot of background information.

As a result, this book is structured to first provide a detailed description of the terrain, and the changes it goes through from season to season.

Next comes the weather, sometimes our friend and sometimes our enemy. Here, too, we examine the seasons and what each one brings.

Then we go to the airplanes, focusing on those that time and experience have proven to be most suitable for our needs.

After that, we examine the special operating procedures required to operate in extremely cold climates.

These thirteen chapters lay the basic foundation necessary for safe bush operations.

Admittedly, this is a lot of ground to cover before getting to the flying techniques, but most of the accidents that happen in the bush can be traced to a lack of knowledge in these areas and the poor judgment that is a co-product of this ignorance.

But that isn't the end of it, for next we need to take a look at the various methods used to find our way around the vast wilderness areas where we will be operating. Both the traditional silk scarf and the new high tech techniques are described, and both techniques have their place in our work.

Finally, we get to explore the flying techniques on wheels and skis, summer and winter, that the pros use. Many readers will want to read this section first, to get a feel for many of the advanced techniques used in professional bush flying.

There is nothing wrong with that. While many books are designed to be read from front to back, this book is one where the reader can skip from section to section at will. But, if you want to get a feel for what it is like to fly in the north, the daily life of a bush pilot, turn to Section III. In that section I decided to push the envelope of textbook writing style. So I integrated much of what the reader will have learned in previous chapters, along with new material, and presented it utilizing the techniques of dialogue, foreshadowing, pacing, and imagery. Here is where the real heart of this book lies, and where the reality of the wilderness lifestyle resides.

I hope you enjoy this book, and find the information of use in your daily flying. As pilots, we are more fortunate than most people, and, in my opinion, nowhere does this show more strongly than when we have integrated our flying with the freedom and beauty of the wilderness.

Section I

Equipment and Environment

Caution—this section contains *advanced techniques* that could be hazardous to nonproficient or poorly-trained pilots. Refer to *Disclaimer and Warning* on page v prior to attempting any maneuvers described in this book.

Section I, Part I

Terrain

The Terrain

The Northern Bush…

There it is, beckoning, calling to us, its miles and miles of mountains and forests, lakes and rivers, tundra and glaciers, beaches and swamps, waiting, and we respond, seeking its vast solitude, its roadless wilderness free of TCAs and ARSAs, smog and traffic, towers and 9-to-5, much as salmon seek their home waters.

A place where we, as pilots, can truly fly. Spread our wings as we were meant to do.

Yet this land, so timeless in aspect, so primeval, is in reality a land of constant change. For not only are there the gross changes that occur between summer and winter, but there are also subtle changes every bit as serious that occur from day to day and often hour to hour. In the summer, glaciers melt, causing rivers and streams to rise and fall, covering and uncovering gravel bars (see illustrations 8 and 9), changing their lengths and textures, silting and shifting channels. In the winter, snow falls, drifts form, and landing areas appear and disappear with an erratic regularity that takes some getting used to. Before we get into the somewhat unconventional techniques (and mindsets) necessary to operate in this environment, we have to develop an understanding of the terrain, the places where we can—and cannot—land.

This knowledge of the terrain is one of the four areas (the other three being judgement, flying technique, and weather) that support our efforts and allow us to practice our dangerous and demanding profession with a reasonable degree of safety.

For, of course, if we fail in any of these areas, we do not survive for long.

Chapter 1

Summer Conditions

Traditionally, summer is the busy season in Alaska and Northern Canada, for the snow has melted and the long days of the midnight sun bless us with 24-hour workdays. Mining, travel, and exploration are in season; it is time for fishing and supply runs for the coming winter. The bush pilot is inundated with requests to fly clients and their gear from the various support centers (Anchorage, Fairbanks, Whitehorse, Kotzebue, Inuvik, Ketchikan, Gulkana, King Salmon, etc.) to almost every spot an airplane can be set down. The main "airports" during this period are:

* For airplanes using wheels: gravel bars, beaches, mountain tops (saddles and ridges) and, occasionally, roads.

* For airplanes using floats: rivers, lakes and coastal waters.

* For airplanes using wheel-skis (when in ski mode): glaciers and high-altitude snow fields.

(As glaciers and snow fields are, in effect, winter environments encountered during summer, I will save them for the chapter on winter conditions and focus my attention here on places where you can land using floats or wheels.)

Water landing areas

Thousands of lakes and rivers, as well as coastal waters suitable for operating floatplanes in Alaska and Northern Canada, provide us with some of our finest landing spots, and, while many places are safe and well-known, it is always wise prior to landing to check for floating or submerged objects which might present a hazard. In particular, you must watch for:

* Large Rocks. The most dangerous ones, because they are so hard to detect, are those just below the surface of the water. In some areas of the Northwest Territories such rocks will be your greatest environmental hazard.

* Floating debris. In the high Arctic this sometimes takes the form of oil barrels abandoned from exploration camps; along the Southwestern Coast of Alaska, expect to find logs and other trash from logging camps.

* Animals. These fellows can sometimes be difficult to see. Once, I almost hit a swimming bear with a Cessna 185. I got the distinct impression he felt I should have been watching where I was landing a bit more closely, and of course he was right.

* Boats. In populated areas this is sometimes a problem. Most Northern pilots consider the worst areas to be Southeastern Alaska near major towns (fishing boats) and the Matanuska Valley in Alaska (speed boats and water skiers following highly erratic tracks).

To these general hazards, you must add the following when using rivers and tidal areas:

* Submerged sand and gravel bars as the water levels rise and fall.

* Areas of shallow water.

An additional problem (which could be considered one of terrain) is brought about by high winds on water. Floatplanes are designed to operate from *sheltered* water, and it is wise not to over-stress them.

For many bush pilots, floats and skis are the preferred way to go, and a surprising number of the old-timers have managed to fly for years without logging any wheel-time.

Gravel bars

For the pilot of a wheelplane, gravel bars are by far the most favored landing areas in the North, and not a few of us would be well content if we could totally confine our operations to them so that, like the floatplane pilot, we never had to use commercial airports. This attitude comes from the calm freedom of gravel bar operations, combined with the often exhilarating sense of being remote from civilization, closely integrated with nature.

Types. There are two types of gravel bars that are used for wheel operations, each quite distinctive:

1) *Non-filling bars.* This type has been made long, smooth, and firm by the action of the water level moving lazily up and down over the years and centuries. Good examples of this geological process can be found below the Knik Glacier near Palmer, as well as in the Spruce Point area where my main base of operations is located (see illustrations 1 through 8 and 10). Here you will find many bars that are not only suitable for the Super Cubs and Cessnas (180s, 185s, 206s) professional bush pilots use, but are even adequate for more citified forms of transport. (I once, on request, operated a Beechcraft Bonanza from gravel bars on both the Chitina and Tana Rivers for photographic purposes, and, while it was far from being as good as a Cessna 180/185 or 206, it still worked out okay.)

2) *Filling bars.* This type of bar is far more interesting from a technical point of view, since it is formed by streams containing large amounts of glacial silt. (From the air these bars look like braided hair or a belt of woven leather.) During the summer, as temperatures rise and fall (often reaching well above 104° F/40° C in the direct rays of the sun), so too does the water level. As it rises it will cut a channel, then fill it partially and move over to cut a new one. This process is sometimes slow, sometimes fast, and can change on an hourly basis during extended hot spells after a winter of heavy snowfall.

In these areas, finding a spot to land within reasonable walking distance of where your customers wish to go can sometimes be quite difficult. Often, a lengthy period of study from the air is required before an area (usually a short, narrow, twisting rocky channel) is located which will be adequate for a Super Cub. Examples in my main area of operations of this type of terrain are the Chitistone and Lakina Creeks.

Length. The ability to judge accurately from the air whether a given landing area is long enough (when conditions are marginal) is a skill that becomes subconscious with time and experience. In Section II, various exercises are given which will assist you in developing this ability, but it is wise to take it slowly, for subconscious skills take time to develop. Starting with long smooth bars, using them regularly, one can, with time, slowly graduate to increasingly shorter and rougher strips, until knowledge and proficiency reach their peak.

Texture. When an inexperienced pilot gets into trouble with an off-airport landing in the bush, it is usually not a matter of a bar's length so much as it is a matter of texture, or just plain unsuitability.

The ideal texture is hard-packed gravel made of stones about the size of a bar of bath soap embedded in sand and smaller stones (see illustrations 1 through 8 and 10 for examples of reasonable textures). Deviations from this ideal are:

* Soft sand. This is easy to perceive. Waves on the surface usually are a sign that the bar is deep and soft. In this type of terrain you will get stopped quickly, and require a long takeoff run to get airborne again. Take care when using it not to apply your brakes very heavily, especially if attempting a short turn-off when there is still a fair amount of forward speed, for this could cause a tire to dig in and put the airplane on its nose.

With its high-flotation tires and lighter weight, the Super Cub naturally does a better job than the Cessna 180/185, though as a general rule both work well if you touch down carefully and keep the ground-run straight so both tires dig in evenly.

And, of course, a deep, soft sand pit can have its advantages. I have a nice little camp near a glacier in the Wrangell Mountains located on a small island where two streams meet, and the landing strip is a 350-foot gravel bar on the north side of the island. At the end of this strip is a line of trees and, due to their height and the approaches, the strip is a classic one-way strip: one must always land to the west and take off to the east.

This would not be a problem for the Super Cub were it not for the prevailing wind. This wind pours off the glacier several miles to the south, hits the island, and boils around and over it, with half going down the runway following the general curve of the land until it meets the other half coming around from the other side of the island. The result: a turbulent 20 to 30 knot crossing tailwind when landing that often turns into a 20 to 30 knot crossing headwind just at the very end, and if I had three windsocks along the length of this strip they would frequently be pointing in wildly different directions.

Fortunately, though, on one side of the island and roughly perpendicular to the main strip is a 150-foot patch of deep, soft sand. This patch of sand faces directly into the wind, and while the approach is wildly turbulent it is operative most of the time.

This allows me to land into the wind on the sand patch, which then requires only a short taxi over passable gullies to the main strip where my tiedowns are located. When it is time to depart, I make my takeoff run down the main strip, holding the Super Cub on as long as possible, then lift into ground-effect and accelerate until I reach the end of the island and the main blast of the wind, which is now a direct crosswind. A tricky area, wind-wise, though all works out rather nicely due to a 350-foot gravel bar and a 150-foot sand pit. It is a classic bush pilot's answer to an interesting terrain problem.

* Rocks. A large percentage of gravel bars are made of material that varies in size from fine sand to rocks the size of a moose's head. As wind and water work on the terrain they will remove the small pieces, leaving behind the larger rocks. Usable ground will vary from reasonably textured sand and gravel with scattered large rocks to bars composed of rocks alone without the binding effects of sand (see illustrations 1 through 9).

How large these rocks can be before the strip becomes too rough for an airplane depends on a number of factors, such as type of airplane, size of tires, current operating weight, wind velocity, and the pilot's willingness to risk damage. As a general rule, rocks over five inches in diameter not embedded in sand and gravel mark the upper boundary for routine Super Cub operations conducted with low-pressure high-flotation tires. Other types of airplanes and tires will require a corresponding reduction in rock size.

Judging the size of rocks from the air is most easily done on sunny days early in the morning or late in the afternoon when sharply-defined shadows are cast. The best bush pilots become as subtly aware of light and its effects as the best nature photographers, and for much the same reasons.

* Gullies. These are an ever-present feature of the gravel bar scene, and can either work for you or against you. While gullies that cross an otherwise nice bar will often ruin it for your purpose, wide gullies in otherwise

unsuitable areas are often the only places where you can land. Gullies fall into three classes:

1) Gullies across a bar that prevent it from being used, no matter what technique is employed. These gullies are either too deep, too sharp-edged, too rocky, or in the wrong place (with reference to the airplane's speed when crossing them on landing or takeoff) to make the bar operational.

2) Gullies across a bar that you can cross without risk during landing and takeoff *if* the proper technique is used. These gullies are generally shallow, and usually far enough from one end to make the strip operational (e.g., a bar 325-feet long with a diagonal gully across it at the 200-foot mark).

3) Gullies you can land in. This type of gully is sometimes wide, sometimes narrow. The bottoms are, on occasion, smooth, though as a rule they tend to be rocky and eroded. Doglegs and "S" turns are common. Gully strips generally require a tailwheel airplane, usually a Super Cub on tundra tires.

(In Section II the techniques required to operate from these strips are described in detail.)

* Driftwood. Driftwood is often found on gravel bars and must be watched for, especially on bars covered by fireweed. A marginal strip can sometimes be considerably improved if you are willing to take the trouble to land lightly loaded on a day with a bit of wind, and go to work with ax and shovel.

Tracks in the sand. It is quite common when flying in areas of extensive gravel bars, or other forms of suitable terrain, to discover airplane tracks on almost any flat surface. This leads to the assumption—especially among Super Cub pilots—that if somebody else landed there the strip *must* be okay. Unfortunately, this assumption has been proven incorrect so many times over the years that it has now become a classic cause of off-airport accidents, and, while the NTSB prefers to relegate the cause to the catchall "Pilot Error" in its accident reports, the reality of the trap that plays "gotcha" with so many inexperienced or unwary pilots cannot be denied.

This trap is unhappily known as the *primrose path* for the following reasons:

* The natural action of the water as it rises and falls can remove a portion of a once-adequate strip while leaving tire marks on the inadequate part that remains.

* The inadvertent repetition of somebody else's mistake can lure the

unsuspecting pilot. Here you have the situation of somebody landing on a gravel bar that was too short or rough, wrecking their airplane, then having it helicoptered out. Left behind are the tire tracks, much like bait for other incautious pilots, tempting them to try and land.

* The skillful actions of an experienced and competent pro who has landed where it is almost impossible *except* under certain carefully defined conditions of temperature and wind in a lightly loaded Super Cub. Later these conditions change (the wind shifts or drops, the temperature rises, etc.), and the strip becomes inoperable once again for even the best of us.

It is wise to avoid primrose paths, understanding that just because there are tracks in the sand does not mean that the strip is suitable for landing.

Gentle fields of purple flowers. Alongside the rivers and gravel bars you sometimes find lush fields of fireweed and other flowers, often ringed by cottonwood and willow. These fields, with vegetation that can easily grow higher than your wheels (see illustrations 11 and 12), often make fine landing areas. Beneath them is dry, well-drained gravel, much like the terrain that lies around them.

However, these beautiful flowers may provide camouflage for the same problems most gravel bars share: large rocks, gullies, and driftwood. Because of this, you have to be quite careful when using a new field whose surface is not well known. It is often wiser to land someplace else (even if it is a few miles away) and hike in for a firsthand look, rather than take the risk of damage to your airplane.

Mud flats

My experience with mud flats (and their propensity to cover a clean, well-maintained airplane with mud) has been happily limited, for they are quite rare in my area of operations.

However, in other parts of the North, especially along the coast, you will frequently encounter this type of terrain. The flats near Anchorage, in fact, are quite popular with duck hunters (as well as airplane repair shops).

Mud flats are obvious from the air and, depending on area and condition, can sometimes be successfully landed on with tundra tires (for this job, the 30" tires work best). There is, however, a bit of risk in trying to use truly marginal mud flats, for you could flip your airplane onto its back. As a result, it is usually preferable (as Mudhole Smith, who started Alaska

Airlines years ago, would sadly have told us) to use a floatplane under such conditions.

Mud flats do not present any problems for a wheelplane when frozen, of course, and during spring and fall when night temperatures drop below freezing these strips are operational during the early hours of the morning.

Beaches

There are many beaches along the coast that provide good landing areas. The same considerations apply to them as apply to gravel bars. Since beaches generally have a fair slope to them, be prepared for the plane to want to head for the sea on both landing and takeoff (a little upslope rudder generally manages to take care of that problem nicely).

"Look at that nice grass-covered meadow..."

The newcomer ("Cheechako" in the vernacular) to Alaska and Northern Canada is often amazed, during his flights across this vast landscape, by all the pretty meadows scattered among the trees. Every once in a while one of them will land on the grass, only to discover that the grass is, in reality, a bog (muskeg). These bogs can be counted on to either take the gear off one's airplane, or flip the plane over onto its back.

Should this happen, there are two ways of getting the airplane out:

1) If the damage is not too severe and field repairs are possible, you can fit the airplane with skis and wait for winter and enough snow to fly it out.

2) If the damage is too severe for field repairs, a helicopter can be hired to haul the airplane out. Naturally, under these conditions, it will have to be partially disassembled.

There are many definitions as to when, and how, a newcomer can lose his or her Cheechako status; perhaps landing in a bog is sufficiently educational to qualify as one of them. It is surely expensive enough.

Mountain landing areas

Some of the most profitable landing areas for the professional bush pilot during the sheep hunting season are the mountain ridges, saddles, and knolls of the North (see illustration 13). While terrain problems are essentially the same as those encountered at lower elevations (texture, length, etc.), there are a number of details to be aware of:

* Optical illusions. The tilt of a strip, whether it is level, slopes to the side, or goes up or down, can sometimes be quite difficult to determine.

(In Chapter 25 Optical illusions are covered in detail.)

* Wind. Like certain gravel bars, mountain landing areas require that the pilot carefully pick his operating conditions. This is not only due to mechanical and convection turbulence, but to the fact that most of the strips are one-way.

As a general rule, the best times to operate from mountain strips are early in the morning and late in the afternoon, with early morning preferred by most pilots.

* Airplane performance. At altitudes of five to six thousand feet (the usual elevation of mountain hunting strips in the Wrangells), you will encounter the following two problems:

1) The engine will not develop full takeoff power (assuming the usual non-turbocharged Super Cub is being used, as is almost invariably the case with wheel mountain landings).

2) The airfoils will not be as effective.

As a result, not only will your landing and takeoff *groundspeed*—though *not* your *IAS*—be higher than what you are accustomed to, but on takeoff your engine will not be producing full power and your prop will not be as efficient. This means you will need a longer than normal ground run for both landing and takeoff.

There are two things you can do to help:

1) Watch your loads. Few of these strips allow gross-weight operations; this is especially important on takeoff.

2) Select landing areas carefully. When the choice presents itself, select an uphill slope. This slope not only helps you to stop within a reasonable distance, but, more importantly, assists you during takeoff where the thin air and lack of power are felt most strongly.

* Tires. Due to altitude and its effects on tire pressure (particularly the low-pressure, high-flotation tires used on Super Cubs), you will usually want to deflate your tires about 20% prior to leaving your low-altitude base camp for your high-altitude spike camp. This enables you to have your tires at the proper inflation for using the rough terrain in the mountains, even though additional care must be taken when using your brakes at the low altitude strip lest you turn a tire on its rim.

(For more on tire pressure, see Part III of this section.)

* Snow. During the fall, fronts moving through the area will sometimes deposit snow on your high-altitude strips. Since these strips are generally marginal for wheelplanes, switching to wheel-skis will not solve the problem. As you watch the weather, keep alternate plans in mind. Alternate plans might include:

1) A lower elevation strip within hiking distance of your high camp.

2) The snow shovel exercise (you can cheerfully airdrop a few to your hunters). This is hardly as bad as it seems, since it does not take all that much effort for a crew of men to clear an area the width of your wheels and about 500 feet in length (the usual size of a mountain strip used by a Super Cub).

3) Exercising patience. This actually works surprisingly well during early fall, since the first few snowfalls rapidly melt off and the wait rarely lasts more than three or four days.

4) Bringing in a helicopter. This is so expensive that all profits from the job will be lost. But, if all else fails…

Roads

For as long as aviation has been established in the North, roads have been used as airstrips, and there are still many villages whose main street serves as their official airport. It has long been the practice in remote rural areas for those who live alongside roads to use them as landing strips, then park their airplanes in their yards.

However, now that the North is developing, this practice is becoming just a memory, for road operations are frowned on by new settlers in most areas.

Still, along the Alaska Highway (which is used as a VFR flyway from Fort St. John to Anchorage and Fairbanks), numerous stretches of the road are suitable for *emergency* use, and I, along with other pilots, have found occasion over the years to use one stretch or another for this purpose.

At present, there is no law (of which I am aware) prohibiting road landings in Alaska or Northern Canada, though a general trend is moving in that direction. I would suggest you use roads as little as possible, and, who knows, maybe the politicos will never get around to outlawing their use (out of sight, out of mind, mode).

Chapter 2

Winter Conditions

My schedule being heavy, I write when I find time, and Christmas morning of 1988 was peculiarly appropriate for the subject at hand. The North was comfortably ensconced under a thick blanket of snow, and the weather showed a spectrum typical for the season from Kotzebue in Western Alaska to Churchill in Eastern Canada. So to set the proper mood for discussing terrain, let's take a look at a few of the reports for 12/25/88:

* *Kotzebue, AK.* Sky 3,500 scattered, 6,500 overcast; visibility 3 miles in blowing snow; temperature 14° F/-10° C; dewpoint 12° F/-11° C; wind 090° at 35 knots, gusting to 40; altimeter 29.54.

* *Fairbanks, AK.* Sky obscured, 4,500 overcast; visibility 3 miles in light snow; temperature 14° F/-10° C; dewpoint 12° F/-11° C; wind 210° at 9 knots; altimeter 29.96.

* *Deadhorse, AK.* Sky 100 obscured; visibility ⅛ mile in blowing snow;

temperature -9° F/-23° C; dewpoint 12° F/-11° C; wind 070° at 34 knots, gusting to 45; altimeter 29.95.

* *Gulkana, AK.* Sky 1,300 overcast; visibility 4 miles in fog and light snow; temperature 2° F/-17° C; dewpoint -5° F/-21° C; wind 360° at 6 knots; altimeter 30.01.

* *Whitehorse, YT.* Sky obscured, 1,500 scattered, 4,000 overcast; visibility 3 miles in light snow; temperature 0° F/-18° C; dewpoint -9° F/-23° C; wind 190° at 13 knots; altimeter 30.00.

* *Yellowknife, NWT.* Sky 2,500 scattered, 6,000 broken; visibility 12 miles in light snow; temperature -6° F/-21° C; dewpoint -13° F/-25° C; wind 140° at 7 knots; altimeter 30.23.

* *Churchill, MAN.* Sky obscured, 900 broken, 22,000 broken; visibility 3 miles in blowing snow and ice crystals; temperature -13° F/-25° C; dewpoint -22° F/-30° C; wind 320° at 17 knots; altimeter 30.31.

Seems about right; no question it's winter. And since weather affects terrain strongly, let's take a look at this Winter Wonderland and see what effect it has had on your landing areas.

Frozen lakes

These are by far the most popular winter landing areas in the North, usable on both skis and, under certain conditions, wheels. Yet, before landing on a lake, there are a number of critical questions that need to be answered, the most important relating to ice thickness.

How thick should the ice be? As with most questions of this nature, there is considerable debate among experienced bush pilots as to minimum safe ice thickness; here I will take, as one always should when dealing with ice, the conservative view. Canadian studies have shown that for fresh water (inland lakes and rivers) at temperatures below 25° F/-4° C, it is advisable to have at least six inches of ice when operating airplanes of the Super Cub class, eight inches with airplanes of the Cessna 180/185 class, and nine inches with airplanes of the De Havilland Beaver class. However, these figures are for skiplanes; when using wheelplanes, or parking over 24 hours, one should have double the listed ice thickness.

When measuring ice from the surface staggered holes are cut along the line of intended operations, and the following values are used:

* Clear ice—full measured value.

* White ice—half measured value.

In other words, if, when cutting through the ice, you discover four inches of white ice and six inches of clear ice, your operational thickness will be eight inches.

How thick is the ice? From the air this can be a difficult question, and there is no real way to tell except from various subtle clues that experience helps us to interpret. The following points should be kept in mind:

* Lake ice is always thickest in the center. If you decide to give marginal ice a try, confine your landing to the central area, and stop there. The airplane should *not* be taxied over to the edge to unload passengers and cargo until the route has been checked from the surface and determined safe.

This checking should be done on snowshoes with the heel-bindings unfastened (so the snowshoes can be jettisoned in an emergency). If you get into thin ice, a curving turn back towards the airplane should be made using a shuffling motion. *Do not stop for any reason,* until back on safe ice, for even thin ice which is brittle will tolerate a moving load far better than a stationary one, and this is where your safety lies.

Under conditions where the center of the lake is safe for the airplane, but the edge-ice is marginal, the airplane should be unloaded where it sits and the cargo transferred to shore by hand or packboard, while wearing snowshoes, when snowmachines or dogsleds are unavailable.

* The areas where streams enter and leave a lake always freeze last, and, when frozen, have the thinnest ice. If stretches of open water extend deep into the lake, the probability is that the ice is thin. I avoid these areas even during the height of winter because they can become covered with a thin layer of ice and then a thick layer of insulating snow, presenting a trap for the unwary even under the best of conditions.

* When using a Super Cub on board skis (see illustration 14) fresh moose tracks become an auspicious sign, for moose have a good sense for ice and if it will support them it will usually support us.

Overflow. An overflow condition causes the snow cover on a frozen lake to become saturated with water. This is usually caused by water backing up in the streams and forcing its way up under the snow, or from warm weather melting the snow on top of the ice and surrounding terrain. An overflow condition can be encountered in temperatures below -40° F/-40° C due to the insulating characteristics of the snow, and this type of overflow is usually found in layers between thin sheets of ice and snow.

An overflow condition can be very serious and could trap you on a lake. One way to detect it from the air is by the wet, dirty appearance the snow will often have. Another method, which is a staple of winter operations, is given in Section II, Part II. Also described in Section II are methods for getting out of this trap, if it does catch you.

(Depth of snow, drifts, and glare-ice will be described separately, for these terrain considerations are not confined to lake environments alone.)

Frozen rivers, creeks, and sloughs

When evaluating the ice of rivers, creeks, and sloughs, you have to take into consideration the underwater currents that can create thin spots and holes. You will not have too much of a problem with holes, for they are easy to detect. Thin ice is *not* easy to detect from the air when the ice is covered with snow, and while its existence can generally be surmised, the clues are often quite subtle. This is another of those situations where only practical, hands-on experience will help (along with plenty of time spent wandering around on the ground studying ice and snow patterns).

Overflow. While this is a problem rivers share with lakes, and must be watched for in exactly the same way, it can also work to your advantage under certain conditions:

* In many places, due to gullies, large rocks, and driftwood, the terrain is far too rough to land on during the summer. During the winter, however, overflow will occasionally cover these areas, then freeze solid, leaving large expanses of smooth ice and snow. These areas then become good airstrips.

* Sometimes, during overflow conditions, the water level only rises enough to fill in the gullies and other low spots. After the overflow freezes, especially on small winding creeks where the wind has blown the surface clear of snow, you will often end up with areas consisting of about 150 feet of ice and 150 feet of smooth gravel. With good approaches this makes a nice airstrip for a Super Cub on wheels or wheel-skis.

Tundra and swamps

There are a lot of soft, wet, marshy areas in the North—remember the lovely grass-covered meadows? In winter, with the ground frozen and enough snow to smooth out the tussocks, these marshes make nice skiplane operating areas.

Arctic pack ice

Pack ice is more interesting than inland lake and river ice, because its highly variable salt content provides a wider range of operating conditions. Here, you must take an even more conservative view towards ice conditions than you did on shore. But, to make up for that is the constantly changing landscape with its strange and awesome beauty, a beauty so seductive, so compelling, that it strikes deep into the heart of any wilderness lover.

The main points to keep in mind are:

* The figures given earlier in relation to minimum safe ice thickness must be doubled for salt water.

* Because snow is often blown off the ice, on occasion color can help in the initial determination of strength; if it is dark (black), it is unsafe.

* While the sight of a moose, or moose tracks crossing a piece of ice, indicates it is probably safe for a light airplane on skis, the same cannot be said concerning polar bear. A polar bear can cross ice that a man on foot cannot.

* Winter ice (this season's ice) has not yet had a chance for the salt to leach out, and snow sticks to it like glue. So will your skis, requiring an additional distance for takeoff, though landings will be satisfyingly short. This ice is moist with a gray or light turquoise color.

* Polar ice is winter ice that has aged for over a year, and is basically salt-free and dry. This allows the snow to be blown free from its surface leaving large areas of polished ice behind. Here your skis will behave as they do on polished lake ice. Its color is dark blue or green-blue.

* River ice will form offshore of large rivers like the Mackenzie. It is also carried for long distances by movements of the pack ice, and there have been reports of it being found over 100 miles from land. This ice is brown, and can be treated like polar ice as far as ski operations go.

Types of snow

The number of different names the Innuit (Eskimo) of Alaska and the NWT have for different types of snow is exceeded only by the number of names the nomadic Bedouin of the Middle East have for different types of camels (or so it is said).

Trying to describe all these types of snow with words (especially in English) is impossible, so only the broad classes are listed below.

* Dry, sticky. This is a product of extremely cold temperatures (-40° F/ -40°C and lower). With this, your skis will have a strong tendency to stick (think glue), and precautions will have to be taken if ground-stays of longer than a few minutes are contemplated.

* Dry, soft. This type, which is often found after a storm (especially if the snowfall was heavy), is comfortable to operate in. But once you have landed, you will find it hard to turn the plane around if you are in a confined area.

* Wet, slushy. This type, a product of spring, is also encountered during mid-winter (during a chinook), or in mild overflow conditions.

* Compact, crusted. Snow settles, the temperatures warm, then go cold, and the snow becomes compact with a hard crust. Walking on snowshoes, you can remain comfortably on top and make good time; without snowshoes, a trip of a hundred yards or so becomes a major exertion. Airplanes will generally break through the crust once they slow down. On takeoff, treat the snow much as you would water and 'get on the step'.

* Wind polished. A hard-packed ski-strip, once wind polished, becomes almost like glare-ice (see illustration 15). Whee...and you better have lots of room to stop.

(See Section II, Part II for techniques to help cope with these problems.)

Depth of snow

Snow depth is important for many reasons, most of them relating to terrain:

* Thin snow. You need a certain depth of snow to prevent damage to your skis. A paved strip (being smooth) requires only a few inches of good compacted snow to be operational. A strip like my home base (see illustrations 16 and 17) requires approximately eight inches of compacted snow, or a good foot of fresh dense snow, to cover the rocks and gravel adequately. An open stretch of muskeg (with high tussocks) could require several feet. There is one strip I can use only when the snow is belly-deep to a moose.

* Deep snow. If the snow is deep, getting stopped will not be much of a

problem, but, in a confined area, getting airborne again might be troublesome. Really deep snow—the type where the horizontal stabilizer is supporting the tail—can require impressively long takeoff-runs.

There are many ways to evaluate snow depth. Most of them are obvious, and after a few seasons, experience will take care of the problem.

(See Section II, Part II, for techniques useful for dealing with deep-snow problems.)

What's under the snow?

This is often the $64,000 question, and there is a long sad list of pilots who have learned the answer the hard way. What we are dealing with here is a process of natural camouflage, and it is very important for the bush pilot to understand how this process works.

As an example, let's consider a gravel bar with large rocks, gullies, and driftwood—mainly uprooted spruce trees from an eroded bank. These trees still have their roots and limbs attached, and are scattered randomly about the bar.

Our story opens during the fall/winter transition, and it has been many days since the leaves turned color, then fell, leaving bare winter branches behind. You have become accustomed to preheating your airplane's engine each morning, and its wing-covers are usually thick with frost. Yet, while there have already been a few snowstorms, none of them has turned out to be serious enough to motivate you into putting your wheel-skis on. Until today, that is, for as you look out the window upon arising you find the snow coming down in a determined manner, with large thick dreamy flakes. Visibility has dropped to less than 300 feet, and not even your Super Cub on the little gravel bar in front of the cabin (see illustration 18) is in sight. It hardly takes an Einstein to figure out what your morning's job is going to be.

The next day it is still snowing, and you are still grounded, but by the third day the front has passed and the sun is out, and upon arising you discover a nice bright early winter day. The world looks fresh and clean, and everything has changed. Cheerfully, you climb into your Super Cub after breakfast and fly out to take a look at the sudden winter wonderland. Upon arriving at your gravel bar you naturally discover it is covered with snow, and looking down in the strong morning sunlight you can easily see where the rocks, gullies, and driftwood lie. The terrain under the snow is easy to interpret, and you would not even consider trying to land there.

Later that day, or the next, a strong post-frontal wind comes up and blows for two or three days. Then once again things calm down, and passing through the area of your gravel bar you notice that it looks somewhat different. The wind has blown the snow around filling in some of the gullies and low areas, and has created drifts in various spots. Now you are no longer able to determine with precision where all the rocks, gullies, and driftwood lie.

Some time later, a few days or a few weeks, another front passes through, and once again it snows a foot or so. Then the post-frontal wind comes up and blows for a few days. This time, when you go out after the weather has cleared you find the area looking rather smooth and level, and if you didn't know it so well you would think a few spots looked safe enough for landing.

Once again the process repeats itself, and this time the area looks fine almost anywhere you look. The danger, of course, is present, and quite real, and it would be easy to snag a ski and badly damage the airplane. If you wanted to land, you would have to exercise caution far beyond what the appearance of the area suggests.

However, once the snow has settled and the drifts hardened, parts of the area *could* be safe for ski operations. The catch is in knowing when this process has reached the proper stage.

Familiarity with an area through the changing seasons is the bush pilot's most important safety factor, after judgment and flying skills.

Drifts

Snowdrifts, which set up almost as hard as concrete, become a prominent part of the landscape as winter progresses, and sometimes it takes only one heavy snowstorm, followed by high winds, to make a landing strip too rough and dangerous to use.

The best way to land on drifts (if the terrain permits) is *parallel* to their general flow, and it helps immeasurably to be able to inspect them during bright sunlight after their formation but prior to landing.

However, often this is not possible after a long flight into a strange area under adverse lighting conditions (overcast, whiteout, blowing snow, etc.), and if you are forced to attempt a landing on drifts that have not been inspected, it is advisable to let down slowly with power and be prepared to lift back into the air if the going proves too rough.

Glare ice

Glare ice is quite different in its degree of slickness from ordinary ice, and it can be dangerous in confined landing areas. It is often formed during winter by a chinook hitting (with high winds and rain), melting the snow to water, then the water refreezing after its passage. Sometimes, especially during spring, you will encounter a cycle where the top surface of the ice melts during the day and refreezes at night. I have experienced this condition when it was so severe—usually when there was a thin sheet of water on the surface and strong gusty winds—that I had to crawl on my hands and knees to my airplane.

This type of ice is easy to detect from the air, especially after one's first (thrilling) experience with it.

Glaciers and mountain snow fields (wheel-skis)

As a basic rule, groups submitting requests for high altitude glacier and mountain landings fall into two distinct classifications:

* Recreational (mountain climbers). Here you can specialize in a small area, for the climbers are interested primarily in Mt. McKinley in Alaska and Mt. Logan (Wrangell-St. Elias Range) in Canada.

* Occupational (scientific expeditions). This is far more interesting, for expeditions can (and do) end up almost anywhere, depending on the requirements of their current mission.

With one exception (crevasses), terrain problems here will be similar to those at lower elevations:

Snow conditions. These will naturally be quite variable, depending greatly on season, altitude, precipitation, temperature, and time of day.

As a general rule, early morning flights will be more productive during weather that has warm days combined with cold nights (below freezing), as the lower temperatures will produce markedly better ski conditions as well as improved airplane performance. And, though not directly related to snow conditions, early in the day the wind will be more benign.

It is not unusual, when the schedule is busy or the terrain demanding, to begin operations as early as 4:00 AM during mid-summer.

Slope. As with wheel operations, a reasonable slope is very desirable during high-altitude ski operations, and for the same reasons.

Length. There are many variables here, and, since this is one of those factors that has the potential for providing interesting moments, it is wise to work up, slowly, to the more critical landing areas.

Crevasses. In some areas these can be your biggest headache, and the fact that they are so often difficult to detect from the air—or even from the ground—does not help matters.

The best thing you can do when going into a new area is to wait until you get a day of bright sunlight and strong shadows, and then study the proposed landing spot for those slight wave forms and other patterns that often reveal crevasses to the educated eye. Needless to say, some pilots are much better at this than others.

Once the initial landing has been made and you are starting to set up camp for your expedition, the strip should be marked out for the duration of your stay. This is done so you can find it easily under adverse conditions of fresh snow, whiteouts, etc. Anything can be used, but I prefer the green-painted bamboo wands that gardeners use, with fluorescent red or orange surveyor's tape tied to the tops. They are light, sturdy, easy to plant, show up well against the snow, and, as a small bonus, demonstrate subtle wind shifts.

Chapter 3

The Transition Times

As winter changes into summer, and summer into winter, you will find yourself in a transitional limbo, a between-time when nature is busily reorganizing your landing areas. Patience is demanded, but so too are your services, for everyone seems to want to get into town or out to some hunting camp, and customer pressures are high. Needless to say, this impatience must be kept under tight control, for if you succumb to its pressures you are not likely to survive for long.

Winter into summer

As the days grow longer, and temperatures rise, landing areas change their nature, sometimes in a straightforward manner, sometimes treacherously. Your main problems will be snow melting on your strips, leaving conditions where neither wheels nor skis can be used; ice breakup on the rivers and lakes; and deep mud (mostly on village strips).

Snow. Under the influence of warm days and spring winds, the snow softens and begins to melt. The rate at which it melts depends on many factors, the most important being density. Hard-packed (compacted) snow will remain much longer than untouched snow. This is as true for your ski tracks as it is for the moose and wolf tracks, now little mounds of ice, that wind their way across the newly-bare ground.

Since few bush strips have snow-removal equipment close at hand, whenever possible during winter I try to leave an untouched area large enough for a Super Cub on each of my strips. In this way, when breakup comes, I will be able to go directly from skis (on the hard-packed snow) to wheels (where the untouched snow has melted).

In areas where drifts occur, the situation is somewhat different, for even though drifts behave the same as compacted snow (which, of course, they are), there is little you can do to control where they end up. As a result, the fluctuating thicknesses mean that once spring arrives you get to contend with bare spots alternating with melting drifts. In this case it is sometimes possible to spend a few hours with a shovel and clear a narrow path (adequate for a Super Cub or Cessna 180/185) through the drifts, thereby opening the strip to wheel operations a few weeks earlier. I have done this for many years at my base of operations at Spruce Point.

If you must attempt a landing on one of these strips before the drifts have completely melted, it is best to use an airplane on wheel-skis and go late in the afternoon when the snow is softest. The wheels will allow you to operate across the bare spots while the skis, in their "up" position, will assist in preventing a nose-over (or loss of control) when the snow is hit.

Lake and river ice. In spring, ice presents you with two main problems:

* Candle ice. This is ice that has gone "rotten". In the spring you get cycles of alternating warm and cold temperatures, and after a while the ice becomes honeycombed with long, thin, vertical, candle-like slivers. Because of these slivers, candle ice can easily be over a foot thick and still give way under you. From the air it has a rough, splotched look (see illustration 19); often, this telltale sign is hidden by the deep fresh snow which is a regular part of the spring weather patterns.

* Thin ice. After a certain point, the ice on the lakes and rivers will become too dangerous to use, and then there is nothing to do except wait it out. This often takes four or five weeks, occasionally longer. Sometimes the ice will go out early enough where the streams enter and leave the lakes

for you to go directly from skiplane operations in the middle (where, if the ice has not candled, it could still be safe) to floatplane operations in the open water.

Overflow. This problem, usually thought of only in relation to lakes and rivers, also occurs in swamps. Overflow operations will be covered in depth in the section on flying techniques (see Chapters 22 & 23).

Glare ice. As mentioned earlier, glare ice is a constant problem during spring; this too will be covered in depth in the section on flying techniques (see Chapters 22 & 23).

Mud. With breakup, many strips in the North turn into a deep sticky quagmire. In fact, it sometimes gets so bad they are difficult to walk across.

Attempting to land on mud, pilots have been known to flip their airplanes. It happens with disconcerting regularity each spring. To alleviate this problem, a number of approaches were tried by different towns and villages over the years, ranging from steel planking to gravel overlays to pavement. The trend now seems to be to give village strips gravel overlays when funds permit, and to pave the strips of major towns.

Still, for many village and bush strips (even those with a gravel overlay), spring breakup is a time of caution, and, as with "rotting" ice, the only course of action is no action until the ground drains and has a chance to dry.

Since night temperatures often drop below freezing during this time of year, important flights can sometimes be made early in the morning (provided the ground is adequately frozen).

Summer into winter

As far as terrain is concerned, this period is not as difficult as breakup, because you will not have to deal with mudholes, drifts, or rotten ice. The problem is more psychological than physical, for this period often overlaps with hunting season, and there the pressure lies. Still, you will have to contend with two problems:

• *Lakes and rivers freezing.* When you can no longer operate on floats, that's it. You must wait for the ice to get thick enough for skis. Be careful. Do not let someone who has a vested interest in your making a flight—your boss, perhaps, who might be more concerned with the company's bottom line than your safety, or an impatient customer—persuade or force you into attempting a landing that you feel is not safe. Too many pilots have died, along with all on board, because of these pressures.

•*Steadily increasing snow depth.* As the snow builds up, you will find yourself needing more room to operate, and soon shorter strips will have to be abandoned until better conditions arrive. Then there is a period, especially on rocky strips, where the strip has too much snow for wheels and not enough for skis. Once again you must exercise your patience.

For landing in fresh snow on wheels, using normal flying techniques, a conservative and accepted figure for maximum depth is one-third the tire diameter. It is wise to stick to this figure as much as possible.

A deep-snow emergency landing technique (wheels)

In an emergency (for example, weather closing in on a VFR flight trapped in a mountain pass) that demands landing *right now* even though the snow on that nice gravel bar or frozen lake is quite deep, here is a technique that could help:

1) Make a slow power-on approach and let the plane settle into the snow *gently.*

2) Do not use brakes.

3) As the snow grabs the plane, it will try to raise the tail; at this point apply more power—*whatever is needed*—to keep it down.

4) Slowly decrease power as the plane slows, *but do not let the tail come up.*

Using this technique, I have landed a Super Cub on 25" tires in two feet of snow. Hardly to be recommended except in a true emergency, it can be done by a good pilot; the problem then becomes getting the plane out. The amount of room this takes is surprising.

(I mention this technique here, rather than in the flying techniques section, because it should be considered an emergency technique, not a normal technique of the working bush pilot.)

Section I, Part II

Weather

The Weather

Descriptions of northern weather go all the way from "It's the worst in the world!" to "It's not so bad." I find myself more inclined towards the latter view, at least away from the cities and coastal areas, for the following reasons:

* It is basically dry in the interior of Alaska and Northern Canada, and precipitation is light.

* During the winter, in cold dry air, icing during instrument meteorological conditions (IMC) is rarely a problem.

* During both summer and winter, cloud levels are usually fairly easy to top even without turbocharged engines (see illustration 20).

* Thunderstorms are mere babies—though still to be treated with respect—compared to the ones further south.

* Except around a few of the larger cities, smog is unknown.

* With generally cold temperatures and low elevations, airplane performance is at its best.

Naturally, there are problems with deep cold, fast moving fronts, high winds, changeable weather patterns, and vast areas without comprehensive weather reports. This is all counterbalanced—once you know the ropes—with the lack of towers, cables, traffic, and pollution, which are endemic down south.

What it finally gets down to is that a little more *individual* thought, responsibility, and technical knowledge is required in the North; in return it is basically a more peaceful and satisfying environment for flight.

Chapter 4

Winter

In an earlier chapter we considered a weather report for various stations across the North, stretching from Kotzebue (on the Chukchi Sea) to Churchill (on Hudson Bay). It was Christmas morning, 1988, and there were extensive areas of scattered to overcast skies, light (or blowing) snow, visibilities running about 3 miles, temperatures as low as -13° F/-25° C, and wind gusting to 45 knots. In the passes, of course, it would have been worse. Not a very nice day for a long trip, and had you wanted to go to Kotzebue or Whitehorse from Spruce Point, you would have had to fly the river valleys—Nizina to Chitina to Copper—VFR until you got to Gulkana (110 nautical miles), then gone the rest of the way IFR. This would have been most comfortable in a wheel-ski equipped Cessna Turbo 206, since the name of the game would have been altitude over the mountains (turbulence, along with the marginal-to-closed passes in snow squalls that, no doubt, would be waiting). (The technique used for flying this type of weather VFR during winter, and what it can lead to, is described in Chapter 24.)

Surely, for VFR traffic, the North that day was living up to its winter image. But is it always that bad?

Well, let's take a look at the weather on January 8, 1989:

* *Kotzebue, AK.* Sky 20,000 scattered; visibility 7 miles; temperature -3° F/-19° C; dewpoint -7° F/-22° C; wind 110° at 15 knots; altimeter 29.54.

* *Fairbanks, AK.* Sky 7,000 scattered, 10,000 scattered; visibility 15 miles; temperature -1° F/-18° C; dewpoint -4° F/-20° C; wind 360° at 3 knots; altimeter 29.59.

* *Deadhorse, AK.* Sky 1,000 overcast; visibility 4 miles in light snow; temperature 18° F/-8° C; dewpoint 13° F/-11° C; wind 090° at 3 knots; altimeter 29.57.

* *Gulkana, AK.* Sky 2,000 scattered, 7,000 scattered; visibility 7 miles in light snow; temperature 0° F/-18° C; dewpoint -4° F/-20° C; wind calm; altimeter 29.53.

* *Whitehorse, YT.* Sky clear; visibility 40 miles; temperature -9° F/-23° C; dewpoint -20° F/-29° C; wind 170° at 13 knots; altimeter 29.67.

* *Yellowknife, NWT.* Sky clear; visibility 15 miles; temperature -33° F/-36° C; dewpoint -42° F/-41° C; wind 080° at 7 knots; altimeter 30.13.

* *Churchill, MAN.* Sky clear; visibility 10 miles in ice crystals; temperature -36° F/-38° C; dewpoint -44° F/-42° C; wind 280° at 10 knots; altimeter 30.13.

Interesting. Only a touch of light snow around Gulkana and Deadhorse, winds gentle and visibilities good. A bit cold in spots. You could, if you wanted, fly from Kotzebue all the way to Churchill VFR (a distance of 2,290 nautical miles via the airways; 1,836 as the raven flies), and by the time you reached Whitehorse you would be, as they say, in "severe clear" weather. And that is something to contemplate, for it is not only mid-winter, but the worst part of it—the December-January period. Admittedly, the temperatures from Yellowknife east are getting near the lower limit of where you may be willing to operate your engine, but since you would be operating in the inversion during cruise this would not present any problems (more on this subject in Section I, Part IV).

So the answer to the question is: No, the weather is not always bad for the VFR pilot. In fact, quite the contrary, for weather moves in huge waves, and when a high settles in, it often sticks around for an extended period. But then, these nice highs bring their own problems, too, as you will shortly see.

Daylight

Nature works in rhythmic patterns, and nothing shows this more than the yearly patterns of daylight in the North, where, above the Arctic Circle,we go from 24 hours of light during mid-summer to 24 hours of dark during mid-winter.

Fortunately, snow cover reflects light very well (see illustration 21), and winter is not as dark as it is often made out to be. With starlight, moonlight, and the dawn-like glow of the sun as it nears the horizon each day to help out, you will have enough light to get by during the dark months.

Naturally, the further below the Arctic Circle you go, the more light you have during winter. At Spruce Point, even during the height of mid-winter, the days are long enough to allow at least six hours of day VFR flying.

Temperatures

Like the seasons and the light, temperatures too move in broad rhythmic waves. They climb and fall in response to their patterns, and thinking of temperature as one would the tides of the ocean disturbed by random gusts is a good way to visualize the process.

A typical short-term mid-winter pattern impressed on the long swell of the seasons, as often observed at my base of operations, goes like this:

* A high-pressure system moves in from the north.
* Temperatures begin to drop.
* Temperatures continue to fall during the night until the coldest point is reached, about an hour *after* first light (cyclic lag), then rise slowly during the day.
* With night, they once again start down, and this time their low point exceeds the previous night's.

This cycle continues, and with a good high-pressure system stabilized in the area the temperature will, after a few days, drop to around -50° F/-46° C where it will remain, with minor fluctuations, for several weeks. (I have seen it on occasion drop below -70° F/-57° C—usually in the month of February.) Finally, a low will begin to move in from the southwest, the sky will start to cloud up, trapping some of the earth's heat, and the temperatures will start to rise once again.

When considering average temperatures, it is necessary to divide the North into specific regions:

* Interior Northern Canada and Interior Alaska. In these areas you will get the lowest temperatures (Snag, located in the Yukon Territory 85 miles from Spruce Point, holds the official record [-81° F/-63° C] for the lowest recorded temperature in North America), and the average for the months of December through February runs around -35° F/-37° C.

* The Northern Coast and Arctic Islands. These areas, while a bit warmer than the interior areas, are more miserable to work in because of the constant wind. A common condition along the North Slope of Alaska is grumpily known as 20/20: 20 knots of wind combined with -20° F/-29° C. While the average temperature in these areas is higher than in the interior, the effect of the wind makes it more serious; the wind-chill factor runs around -68° F/-56° C.

* Southeastern Alaska and the Alaska Peninsula. In these areas it rarely gets much below freezing because of strong maritime influences, but the high winds and incessant rain and fog make it more uncomfortable than either the interior or the northern coastal and island areas.

Temperature inversion

In your elementary weather studies, you learned the lapse rate, the normal decrease of temperature with altitude; in the troposphere, it averages 2° C per 1,000 feet. However, under certain conditions, *this situation is reversed,* and temperature increases with altitude: this is called an inversion.

An inversion has both a negative and a positive effect on flying.

* Assume a very dry surface temperature of -30° F/-34° C with a moist layer of air a few thousand feet up. Climbing through this will cause your airplane to frost over completely, and the view from your window will be the same as in cloud. Until this frost burns off, or until your defroster thaws a clear spot in your windshield, the flight will have to be conducted on instruments.

For the VFR pilot, the only thing that can be done in a situation like this is to open the side window. Admittedly, this will make things a bit chilly in the cockpit, but surely this is better than losing control of the airplane. After all, in twenty or thirty minutes the frost will burn off and everything will once again be all right.

This effect, in clear air, is happily quite rare. In years of flying it has only happened to me a few times.

* Really low surface temperatures (-50° F/-46° C, and below), combined with a pronounced inversion, make it possible to take off, climb into the warmer air, and complete the flight there. However, your descent and approach to landing will have to be planned carefully, and that subject will be covered in Section I, Part IV.

Needless to say, this flight technique should be used only in an emergency, for not only is it hard on the engine and airframe (especially during takeoff and landing), but should a problem develop while en route that causes you to end up on the ground in a wilderness survival situation (for example, a prop-seal or oil line failure), the extreme cold could be quite deadly.

When conducting *normal* operations in cold weather (above -30° F/-34° C), it is also advisable to seek out and use the temperature inversion whenever possible. This makes the cabin heater more effective, and it is easier on the engine. This is one reason why so many Super Cubs and other VFR bush planes have household-style thermometers taped to their wing struts.

Cloud cover

During much of the winter the weather is clear and cold. The cloud patterns are cirrus, then scattered to broken, leading to overcast (see illustration 19), then (after the front moves through) clear once again. The exception to this rule is along the southern coasts and the Panhandle, where maritime effects keep things cloudy and wet much of the time, and where ice can more readily be found.

Whiteout

When scattered to broken clouds finally become overcast, and the overcast is thick and uniform above snow-covered terrain, the visibility condition that develops is known as a whiteout. The parallel rays of light from the sun become broken up and diffused as they pass through the overcast. Then, hitting the snow from many different angles, they reflect back to the overcast, which once again reflects them to the snow, which reflects them back to the overcast in a ceaseless cycle which eventually eliminates all shadows. This is a situation similar to, but opposite from, complete dark—complete white. Flying in a whiteout has been likened to flying in a bottle of milk. I equate it with flying in cloud or fog.

Whatever the analogy, it is a dangerous condition. On the ground, you could walk off a cliff without seeing it. Walking down to my airstrip from my cabin at Spruce Point—a distance of half a mile—I often fall off one side or the other of the snowshoe trail, and dark objects (brush, trees, the spruce boughs lining my airstrip) seem to float eerily in midair. From the cockpit, once airborne, there are no visual references, and flying is strictly an IFR operation (though legally VFR), unless operating below the timberline and wind has blown the snow off the trees. Under these circumstances VFR flight is possible so long as you do not cross any large lakes or other snowfields.

Ice fog

Ice fog is exactly what its name implies—fog made of ice crystals.

Under mild conditions, the degree to which it will obstruct vision depends on which direction you are facing in relation to the sun. If you are facing into it during early morning or late afternoon, it might completely blind you. If you are facing away from it, you might be able to see for miles (see illustration 22). The crystals sparkle and shine like exotic but deadly jewels, and the ice fog glows with a strange cloud-like luminosity.

Under severe conditions, ice fog is another matter entirely. It is every bit as serious as regular fog, except for the possibility of structural icing, and you can encounter it in the bush as well as the cities. Fortunately, it is usually a local problem confined to a specific area.

The basic requirements for ice fog are:

1) Still, very cold air capped with a pronounced temperature inversion.

2) A source of moisture.

Ice fog can be separated into two classes:

City conditions. The sources of the triggering moisture in the cities are cars, buildings and the airport. Most cars are left running when it is very cold to keep them warm while parked, and buildings pump their fumes (oil, coal, wood, gas) into the air in thick copious clouds. At the airport the jets and other traffic keep things going nicely, and the colder it gets the worse it gets. These conditions affect VFR and IFR pilots differently.

* VFR. This section could be titled *Trying to Cope (Psychological Department),* for, to a VFR pilot, a layer of ice fog blanketing the city and trapping him on the ground can be a maddening situation—especially when visibilities have been drifting between 0/0 and ILS minimums for days. Hardly helping matters is the knowledge (provided by IFR pilots to FSS on departure) that (1) the top of the layer is around 1,200 feet AGL, and (2) a mile or so beyond the city there is unlimited visibility under a clear blue sky. Obviously, this sort of predicament has its dangers, and the longer it goes on the greater the pressures become.

I should know, for over the years I have experienced this situation a number of times with VFR-only airplanes, and am more than familiar with that helpless feeling which eventually transforms itself—in its healthier form—into a compulsive desire to remove the wings from the airplane, truck the whole thing out of town a few miles where it is clear, replace them, and depart via the highway. This feeling usually comes around the 17th day one has had breakfast in the hotel coffee shop while moodily staring out the window at the cars, trucks, and pedestrians groping their way down the cold, foggy street.

On a short-term basis—for those of you who may be reading this while waiting out ice fog in some lonely city far from home—there is little I can provide in the way of advice except to counsel patience and remind you that sooner or later a warm front will show up with welcome blizzards and associated phenomena to break the trend of the CAVU weather that has set the stage for this latest demonstration of man-and-nature-in-the-North. In any event, I hope you will never have occasion to challenge the record I set back in the early 60s when I was stuck amongst the amenities of Ft. Nelson, BC, with a VFR-only Super Cub for 29 days...

* IFR. On a long-term basis—since ice fog hardly exists as an operational problem to the pro as long as he has the proper equipment (IFR airplane and airports) to deal with it—I recommend that for city operations the bush pilot forget about VFR and acquire at least the minimum equipment (radios and instruments) that will permit dependable IFR departures and arrivals.

Bush conditions. While ice fog is far less of a problem in the bush than in the city, you can occasionally create it by the operation of your airplane when conditions are just right. The triggering mechanisms will be your engine exhaust and your condensation trail. Of the two, engine exhaust is the more critical.

* Engine exhaust. On very cold but clear days with a narrow temperature/dew point spread, you can sometimes close your strip by simply starting your engine, warming it up, then applying enough power to taxi into position for takeoff.

When this occurs there is little you can do except continue to the end of the runway, stop the engine, put the engine cover in place (see Section I, Part IV), and wait for the ice fog to dissipate. This usually takes about 20 minutes.

* Condensation trail. This problem shows up when you are inspecting a prospective landing area from the air. Your pass could lay down a trail of ice fog that could close the strip for a period of time.

Fortunately, this situation rarely gets so severe that you cannot get away with the minimum two passes required to drag your skis through the snow to prepare the surface and check snow conditions (see Chapter 23). And, since the ice fog will have a tendency to form thickest at the far end of the strip where you add the majority of your engine exhaust to the condensation trail (when applying go-around power), this should be taken into consideration, along with any light wind that may cause the ice fog to drift into or out of your path. This last point is especially important on lake landings where you often have a wide choice of landing directions.

Steam fog

Steam fog is caused when *comparatively* warm water from open leads in the pack ice or open spots on rivers, inlets, and lakes is exposed to cold air.

While steam fog is usually local in nature, it can help trigger fog over vast areas, and I have seen whole inland valleys go from CAVU to 0/0 in about an hour after sunrise (see Chapter 24).

Steam fog, unlike ice fog, will cause light rime ice to form on the airplane as one passes through it, and this should be kept in mind when making instrument approaches and departures. (It is also one of the few times the windshield anti-ice system available for the Cessna 206 will prove worth its cost.)

Advection fog

This is a problem of the coastal regions and is caused by comparatively warm, moist, oceanic air flowing over the cold, snow-covered land. If there are mountains, lifting will cause low stratus to form, as well. This fog dissipates rapidly as you go inland and is rarely seen on the lee side of the mountains. Be warned, though: rime ice may form on aircraft, and can become severe. Advection fog often sticks around for days, sometimes weeks, which is one of the reasons I prefer the interior to the coast.

Upslope fog

Upslope fog results when moist, stable air is cooled adiabatically as it is moved up sloping terrain. As long as the upslope wind continues, so too will the fog.

This condition can be encountered along the Alaska Highway in the Ft. Nelson area, and should be watched for. It can be persistent, cover vast areas, and reach high altitudes.

Fast-moving weather systems

The North is famous—perhaps I should say infamous—for its fast-moving weather systems and frontal activity, and this activity reaches a peak during winter. As a result, a careful evaluation of all available weather information should be made prior to each flight, and kept up to date while en route.

A few of the major problems associated with these weather systems are:

* Snow squalls. While snow squalls are almost meaningless to the instrument pilot *on a route where he can fly IFR,* they can be very dangerous to the VFR-only pilot. They form and dissipate rapidly and are especially serious when combined with high winds and turbulence in a long narrow mountain pass.

The best VFR technique for flying snow squalls in remote areas with a skiplane, if inadvertently caught by unforecast weather, is to lake-hop, going from landing area to landing area and circling when necessary so that you never risk getting caught away from a place where you can land and wait out the bad spells. This technique works surprisingly well as long as you have the self-discipline required to keep from pushing the weather too far and are willing to spend the night camping out, if necessary (see Chapter 24).

* Widespread snow (moderate to heavy fall). Alaska often gets widespread conditions of light to heavy snow, and, as our December 25, 1988 weather report showed, it can extend for impressive distances. For the IFR pilot on a long trip (if his route allows instruments), this can be one of the nicest "bad weather" conditions available. Calm and dreamy and peaceful, the world becomes a very private place.

This type of snow is called sublimation, for it forms directly from water vapor without cloud formation. Therefore, airframe ice is quite rare.

However, there is one potential problem you need to guard against. Under conditions of heavy, wet snow (usually occurring in spring), there is a slight possibility your air inlets and engine induction air filter could become plugged. While this has never happened to me, I have heard of it happening to others. Careful engine monitoring is called for, and if manifold pressure begins to drop a precautionary landing is indicated.

* Snow falling from, or through, clouds. If there are clouds (low stratus, stratocumulus, nimbostratus) associated with the snow, you have to be careful of them, for they will contain ice. Pilots *with the proper equipment* handle this situation by climbing through the cloud deck to a area between the layers, and proceeding from there. For, of course, unless you are actually in cloud (except for freezing rain or certain wet snow/sleet conditions), you will not get ice.

* Freezing rain/ice storms. Freezing rain is caused by warm fronts and is a subject of some controversy. Some pilots shrug and say you can climb out of it (after all, is there not warm air above?); others prefer to wait these conditions out, with their airplane, if possible, safely ensconced in a heated hangar. I belong to the latter group, unless I happen to be flying a jet with hot wings and cowl heat, but for some reason jets do not really work out all that well on skis in a bush operation.

* High surface winds. Pressure gradients can create very high winds, especially in and near mountain passes. These wind-storms, which often come up suddenly and sometimes exceed 100 knots in the gusts, can last for days. When this happens, there is nothing to do but make sure your airplane is well tied down as you take a well-deserved vacation from work, and become reacquainted with your family.

However, most of the time the wind stays below 50 knots and so is manageable. I prefer the following limits for gusts for *ordinary* operations: 30 knots in the Super Cub, 40 knots in the Cessna 180/185, and 45 knots in the Cessna 206. Emergency conditions, where one will have ground help to hold the wings once on the ground and while unloading, are a different matter (see Section II for more on this subject).

* Blowing snow. Visibility in blowing snow can often be a serious problem after a major storm and can easily approach the 0/0 stage on the surface. (Remember the SA for Deadhorse, where, on December 25, 1988, they were reporting 100 obscured and ⅛ mile in blowing snow?) As with ice or steam fog, you must climb through blowing snow to "on-top" during takeoff; on landing, you will often find you have to abort and go to your alternate, where wind or snow conditions will be less critical. For this reason, it is wise to carry plenty of fuel and to keep abreast of current conditions along and to both sides of your route. As the instrument pilot always has an alternate available in case of need, so too does the pro bush pilot: neither cares to fly without a good "Plan B" waiting in the background as insurance.

Turbulence (mechanical)

Mechanical turbulence can get bad at times, especially in and downwind of mountains. If you can see snow blowing off the peaks it is a clear sign of moderate to severe turbulence within 3,000 to 5,000 feet of high terrain, on the downwind sides of peaks and ridges, and through the passes.

However, depending on the stage of the windstorm, you can often avoid a rough ride by flying either down low or up high rather than in the more usual middle altitudes.

* Down low. During the initial development of a windstorm, before the winds reach the surface, the lower altitudes will provide a smooth ride if you remain within a few hundred feet of the surface. Then, as the winds lower and pick up speed, a distinctive form of turbulence will begin to be felt. This turbulence, described as *edgy* or *nervous* (and having a pronounced rodeo-like connotation), contains considerable shear and is an indicator of future events, even though the wind on the surface may still be calm. Under such conditions you should not make a long round-robin flight without considering the possibility you might get caught in turbulence severe enough to force you to land someplace where your airplane could be

tied down and the storm waited out. This "waiting out the storm" might take three or four days, if not longer, and if a tent in the bush were your temporary home, things could be quite grim.

* Up high. Sometimes an area of strong winds extending from the surface on up with associated turbulence will lie over part of your route, and, if so, it pays to check out the higher altitudes. I have often flown above lenticulars and received a good ride while everyone below was filling the airways with loud moans. I remember one flight nonstop from Whitehorse to Anchorage that was conducted at 16,000 feet MSL. The dust on the surface was blowing in impressive clouds. A Navajo en route between Fairbanks and Whitehorse at 4,000 feet AGL notified Northway Radio, just as we were crossing overhead, that they were going to make an emergency landing at Tanacross due to turbulence and airsick passengers. Gulkana FSS reported a helicopter had just been forced down by wind and severe turbulence at the mouth of Isabel Pass (between Gulkana and Big Delta). And we sat there, oxygen masks on, in silky-smooth air, making our position and pilot reports as each station *slowly* passed under our wings. (This is one of the reasons I fly a turbocharged, rather than a nonturbocharged, Cessna 206, even though the operating costs of the Turbo 206 are considerably higher).

However, there is one thing to remember when flying in smooth air at high altitude: the descent will have to be made through rough air, and this will require careful planning to keep out of the rotor on the downwind side of the mountains.

Altimeter errors

Because you will be doing a lot of instrument and night flying during this time of year you will need to keep the following in mind:

* Temperature. Cold temperatures will cause your altimeter to read incorrectly, in the dangerous direction. At 10,000 feet MSL, if you were using an altimeter setting furnished by a sea level station during a cold snap, your altimeter might be off by as much as 1,000 feet were the outside air temperatures in the neighborhood of -25° F/-32° C—a situation that is not at all uncommon during winter in the North.

This temperature error can also cause problems on an instrument approach in cold weather, and can be especially serious when using a non-precision approach into a sea level airport. (Jeppesen, in their Alaska/Canada approach chart service, provides an altitude correction chart that must be used for cold weather approaches when using Canadian airports. This page [CA-19 as this is being written] is well worth studying prior to cold weather operations.)

These errors can be calculated. I use a Jeppesen Pro Star navigation computer, but you can use the standard rule of thumb that states that for every 20° F (11° C) temperature varies from standard there is a 4% error in altitude. However, the problem with these calculations is that they are only *approximations,* because the formula assumes that pressure and temperature lapse rates are standard, or, if not, that they are at least uniform. Therefore, you can't use them for *absolute* minimum obstruction clearance altitudes, but only to set a ballpark margin.

* Pressure (en route). You probably learned the little mnemonic, "From a high to a low, watch out below," in flight school days, and though it is trite, it's still relevant, especially in the North with its fast-moving fronts, rapid pressure changes, and widely-spaced reporting stations.

* Pressure (surface). Just to introduce a bit of variety, nature every once in a while tosses you a wild card: an intense high pressure system that raises the barometric pressure above 31 inches. Since pressure this high takes you beyond the maximum barometric setting capabilities of your altimeter, it not only restricts you to VFR landings during the day, it also precludes night operations unless the destination airport is VFR *and* has either an ILS, PAR, or VASI available.

* Wind effect over mountains. This is the cute one. High winds cause the pressure to drop when flowing over mountains or through passes (venturi effect). Once again your altimeter will be in error, sometimes, under extreme conditions, by as much as 1,000 feet.

* Station error. Though not truly a station error, it can have the same effect. What happens is that you get a deep trough of low pressure lying over an area of high pressure. You take off with the altimeter setting valid for the surface, fly into the low pressure, then radio ahead and receive another altimeter setting, also from a station with high pressure. This leads you down the primrose path (there are many of them in bush flying), unless you have studied the actual pressure gradients along your route (e.g., the 850, 700, and 500 millibar charts).

* The domino theory. All these problems can combine, in various

patterns, and the best rule of thumb is to add 2,000 feet to the official MEA during night and instrument operations over mountains.

Chapter 5

Spring

After the long cold winter finally comes spring, with its steadily increasing daylight hours, rising temperatures, wind, wet snow, low cloud decks, icing conditions, and beautiful calm sunny days (see illustration 2 for an example of spring sun bathing, Alaska style).

Steam fog

Steam fog, which posed intermittent problems during the winter, can be a daily problem along the Arctic Coast during this time as the pack ice begins to draw away from shore.

Since this fog is generally local in nature, with tops around 1,200 feet AGL, and rarely comes further inland than a few miles (unless there's an offshore wind), alternate airports to consider might include village strips, gravel bars, and other areas where you could land within radio range of your destination to wait for the weather to rise above IFR minimums, or, for VFR pilots coming in from the bush, to turn VFR.

Snow

The occasional wet heavy snowfalls you had to deal with during winter chinooks are common during this period, and you will have to take the same precautions now as you did then.

Snow squalls

Snow squalls will occur, sometimes turning into freezing rain. Passes will be blocked, and patience once again will be much in demand.

Icing conditions

Cloud decks will be low with temperatures close to freezing at ground level, and ice will become a problem. Instruments will not be the cure-all they were during winter, and, like the VFR pilot, you too will have to exercise patience.

Beautiful calm sunny days

Call in "well" and take a day off with the family. The boss won't expect you to fly when it's nice; after all, aren't the hard jobs in tough weather always reserved for you?

Chapter 6

Summer

Except for fast-moving weather systems, the problems of summer are not much different in the North than the South. This makes summer the best season for a first visit by those who live in the Lower 48, and the period between June 1st and August 15th will provide the optimum combination of warm weather and good flying conditions. Keep in mind that since weather systems *are* fast-moving, waiting out weather problems works very well and is recommended.

Chapter 7

Fall

Fall is one of the busiest and most profitable seasons for the professional bush pilot. There are sheep, moose, and bear hunters wanting transportation to and from their hunting camps; miners and summer scientists needing to close down their mines and scientific outposts until spring; and trappers needing to get supplies in for the long winter.

The season also brings steadily shortening days, along with low clouds, snow squalls, freezing rain, and high winds.

In dealing with these problems, use the techniques described earlier.

It should also be noted that some of the nicest flying weather comes during this time of year. You can expect an Indian Summer during September that lasts from two to three weeks, bringing brisk mornings and clear calm sunny days. Then it is suddenly over, the weather changes, winter arrives, and stillness, like the now-falling snow, settles over the land.

Section I, Part III

The Airplanes

The Airplanes

Well, it's time to let the cat out of the bag, even if it does make a few pilots unhappy, so here goes. Successful bush flying has always been more a matter of one's level of awareness and skill, based upon knowledge, experience, and judgment, than the specific type of airplane (or equipment) one operates. For this reason the pro can, and does, accomplish "miracles" with almost any of the light airplanes available today. I once used a Beechcraft Bonanza for off-airport operations. And I know of a guide who sometimes uses an antique Beechcraft Staggerwing to transport his customers from Anchorage to his main lodge, where he then dispenses with romance by switching to the prosaic but far more practical Super Cub. The Alaskan air taxi operators successfully use the tricycle gear Cessna 206, 207, and Piper Cherokee Six 300 in their daily operations to the various Native villages, whose (mostly) well-maintained dirt and gravel strips also see a number of light twins like the Aztec, Baron, and Cessna 310.

However, for the pro, whatever advantages these different airplanes might possess in the way of charisma, passenger comfort, speed, ease of loading, or customer acceptance (the larger the airplane, the more engines it has, the happier they are, no matter what the job actually demands), are counterbalanced with quite serious disadvantages with respect to their utility for off-airport operations. Unfortunately, once we get away from the nice village strips, and long, smooth, hard-packed gravel bars (which I used for the Bonanza), the problems begin to mount.

At this point in the game, let's take a hard-nosed look at priorities. On the practical level, these include, but are not limited to:

The economic considerations. This subject is beyond the scope of this book with the following exception:

The bush pilot's trade is one of providing economical rather than extravagant (i.e., helicopter) transportation into remote areas. Therefore, anything that might restrict where he can land will also restrict his ability to compete. For this reason, minor differences between airplanes—e.g., a degree of additional comfort versus a degree of additional STOL capability; a degree of additional range or load versus superior handling qualities—will have to be taken into consideration when it is time to make an equipment decision. As a result of this, to use the example of two popular competitive airplanes, the pro will always choose the Piper Super Cub over the Citabria Scout, even though the Scout can go *almost* anywhere the Super Cub can, and has a number of minor advantages over it.

Wheels and skis. Because off-airport landings and takeoffs often involve areas of difficult terrain that can take airplanes to their maximum structural, performance, and longitudinal stability limits, you must utilize "conventional" landing gear for most of your serious jobs in the deep bush.

While there are a number of important technical reasons behind this necessity, including the forces acting on the horizontal tail surfaces at rotation, and control of the wing's angle of attack during the takeoff roll over rough terrain, the two main ones are:

* Structural and maintenance considerations. The "toe-stubbing" weakness of nosegear assemblies is well known, and while tricycle gear has its advantages, especially in very high winds on paved runways, the loss of payload you would have to suffer were they made adequately strong for rough-field operations is unacceptable in anything smaller than a Cessna Caravan or De Havilland DHC-6 Twin Otter (equipped with high-flotation tires on all three wheels). Also, the prop on a tricycle gear airplane is much closer to the ground than with a tailwheel airplane, and therefore is much more susceptible to gravel damage during takeoff.

* Longitudinal stability. Difficult terrain demands the use of a landing gear geometry which will best handle surface conditions. The simplest way to achieve this is to put as much distance as possible between the main gear and the third point of support, while keeping the center-of-gravity *behind* the main gear; this is especially important in ski operations. Conventional gear does this.

Floats. The list of airplanes certified for floats is limited, and those who operate from water must make their selection from this list. Fortunately, every airplane that attracts the pro has this certification. There are also a number of airplanes that, while not suitable for pro bush work, are suitable for the sportsman pilot.

With regard to single-engine amphibians, none of the "flying boats" have managed to work out for the bush operator, though they seem to be satisfactory for recreational use. For his amphibious needs, the pro usually restricts his selection to the Cessna 185, Cessna 206 (I prefer the turbocharged version), and the De Havilland DHC-2 Beaver. However, keep in mind that amphibious floats exact a considerable weight, performance, maintenance, and cost penalty over straight floats, which greatly reduces their utility for commercial bush operations.

High-wing or low-wing? When flying operations take you from airport to airport and do not require you to land anywhere else, this consideration can be decided on the basis of aesthetics, ride quality in turbulence, handling characteristics under high surface wind conditions, etc. Unfortunately, in the bush you do not have the luxury of this choice, for your operational requirements unquestionably demand a high-wing airplane so you can adequately clear brush, stumps, deep snow, high drifts, ground swells, etc.

There are, of course, numerous other considerations involved, but most of these are individual requirements so I will not go into them here. Suffice it to say that over the years, among the pros, both in the bush and air taxi ends of the business, there has been an equipment shakedown based on the various *practical* considerations of the trade. The end result has been that certain airplanes have proven themselves to be the best ones for the job. There is little overlapping among them as regards job function, and if I were to classify them in automotive terms—a sacrilege, many pilots would say—the list would go like this:

* Piper Super Cub = Jeep
* Cessna 180 = ½ ton 4x4 pickup truck
* Cessna 185 = ¾ ton 4x4 pickup truck
* Cessna 206 = Small passenger/cargo van
* Piper Cherokee Six 300 = Small passenger/cargo van
* DHC-2 Beaver = 1 ton stake-bed truck
* Cessna Caravan = 1¾ ton moving van

In this book, depending, of course, on the job at hand, I will call upon the Super Cub, the Cessna 180/185, or the Cessna 206 for my examples. The reason for this is because these airplanes are the ones most often called upon by experienced pros. As for the others, the fine Beaver, used mostly on floats in Northern Canada, is getting a bit long in the tooth and expensive to operate; the Piper Cherokee Six 300 is too restricted—it is at its best when hauling mail and groceries from town to village, or landing on the beaches of the Alaska Peninsula; and the costly Cessna Caravan, with its Category II instrument capabilities (so needed during the long ice-fog winters), is not yet in wide use in the North.

Chapter 8

The Super Cub

In many respects, experience is the best teacher. For the bush pilot, experience teaches that when the job requires a risky landing on some rough gravel bar, short mountain ridge, tiny pot-hole lake, or postage-stamp sized chunk of pack ice, the Super Cub is in a class by itself.

It is hard to say exactly what it is about this airplane that makes it better than its competitors, but a lot of it seems to be in the combination of stalling characteristics, control balance, and the way its stiff, sturdy landing gear, equipped with low-pressure tundra tires, handles difficult terrain. Most bush pilots appreciate its agility, not only during approach but on the ground as well. For some reason, the flow of forces seems just right for this type of work.

Its construction is also uncannily suited for work in remote areas. With its simple steel frame covered with Ceconite fabric (easily patched when necessary) and dependable 2,000 hour TBO Lycoming engine, the Super

Cub can be operated for extended periods far from maintenance centers and parts houses.

However, for those who fly the remote bush professionally and operate out of the most marginal of strips—often hardly more than 200 measured feet or so of rough, one-way gravel bar—the stock Super Cub, as good as it is, is not completely satisfactory. It requires certain modifications to achieve its full potential. Let's explore the modifications professional bush pilots require for Cubs operated on wheels and skis in Alaska and Canada.

Prop. The stock prop works fine for photography, pipeline patrol, and running around the "spread" down in Texas, but for the bush pilot it is too short and has too coarse a pitch for optimum takeoff and climb. Prior to 1967, the bush pilot's only option was to have the stock prop repitched at the local prop shop. Then Rodger M. Borer designed and obtained an STC for a long McCauley prop.

The Borer prop, especially in 42 pitch, makes a decided difference in takeoff and climb, and few Super Cubs in Alaska or Canada today are without one. I recommend it most highly.

Safety-cables. The rough fields which we use in our daily work place great demands on our landing gear. While much can be done to alleviate this problem with proper tire selection and inflation (see below), and flying technique (see Section II), there is little that can be done about rough, rutted ski strips, especially those in small villages and next to lodges that get used a lot yet receive no winter maintenance.

Recognizing that the main problem was either the upper inside attach-points (ears), which tend to break at the fuselage, or a failure of the shock assembly, F. Atlee Dodge (the grand master of Super Cub mods) came out with a set of safety-cables that provide a backup (see illustration 14). Truly good insurance, especially on skiplanes, they have been credited with saving many a prop and wingtip from serious damage, and it is rare to find a bush-equipped Cub without them.

The best way to install these cables is to mount them so the lower cables ride on top of—rather than hang below—the shocks (see illustration 23). This helps to prevent your landing gear from taking on the appearance of a florist's display after a visit to some nice field covered with fireweed during the fall hunting season.

Axles. The landing gear of the Super Cub has hollow axles, another weak point when the airplane is operated on rough terrain (especially on skis). The standard treatment is to have your mechanic insert short pieces of galvanized steel water-pipe into the axles for reinforcement; once this is done, the problem is taken care of and can be forgotten.

Tires. There are two good reasons for putting large tires on airplanes that must be operated on soft, rough strips, only one of which is well understood.

1) Footprint. If the field is soft, it is obvious that a larger tire will increase flotation, and this in turn will decrease the risk of damage to the airplane when landing. Depending on where and how you must operate, sizes from 8:50x6 up to the 30-inch tundra tire will prove satisfactory.

2) Cushioning. When you land on a rocky gravel bar, or have to go across small gullies, you need something to relieve the shock, and low-pressure tires will do this. The tires roll over the rock (or whatever) the way a pillow might, enfolding it gently. Yet, oddly enough, many Super Cub pilots do not seem to understand this, for most of them carry far too much air in their tires.

For years I have used approximately six pounds of air in my tires, and while they always appear half flat (note where the dirt-line is on the tires in illustrations 6 and 23), they permit very soft and gentle landings and takeoffs from truly rough terrain. The old 24x10x4 and the currently available 26x10.5x6 tire sizes are the ones most bush pilots in Alaska prefer. They seem to provide the best balance between footprint and shock-absorption for general use of any of the tires currently available.

(A note on listed tire sizes: Over the years there have been a number of different manufacturers of tundra tires, and each seemed to give them a slightly different size. Back when I started flying the bush, the only tundra tire available for the Super Cub was the 24x10x4 [see illustration 23]. Later came a real nice 25x10x6 4-ply tubeless tire [see illustration 6]; and now we have the 26x10.5x6 6-ply tire which uses a tube. In this book, when I mention a 24, 25 or 26 inch tundra or high-flotation tire in an operational sense, I am referring to the 25-inch category, for the differences, except for the superior 4-ply, are not relevant.)

As for the 30-inch low-pressure tundra tires, except for a few special jobs, e.g., when operating from mud flats or certain tundra areas, their size works against them. They require noticeably longer ground-runs, in comparison with the 25-inch tires, due to (1) very poor braking on landing, and (2) added friction and drag during takeoff. This, combined with the added weight, makes them a poor choice unless absolutely necessary.

The 8:50x6 tires (whose main benefit is cost) are passable on terrain that is not too rough or soft, but should be avoided whenever possible (see illustrations 5 and 28).

Windshield defroster. F. Atlee Dodge makes a kit for this, necessary during winter if you want to see outside the window.

Metal headlining and extended baggage compartment. Another of Atlee's modification kits, this markedly increases the cargo space in the Super Cub.

Tailspring. The stock factory springs will not hold up under the job, and must be replaced with heavy-duty springs. For many years I had them made up by a local spring shop in Anchorage, but now Piper offers a set for their ag-model that does the trick. The Piper parts numbers for this set are #64079-02, #64079-03, and #64079-04; one of each is required, and a small bottom-plate will have to be made, of hard steel, to hold them at the rear attach-point, using long bolts down each side of the springs.

"STOL" mods. One of the Super Cub's strongest points is the exquisite harmony of its control interrelations combined with the way its stall progresses. This allows the experienced bush pilot who has a good feel for his airplane's slow-flight characteristics to make the final part of his approach into a short strip with the wing right on the edge of the stall (see Chapter 17). However, when you alter the Cub's wing from factory specifications by drooping the ailerons or wing-tips, or the leading edge—all common STOL mods—so as to slightly increase the angle-of-attack the wing can achieve, thereby slightly lowering your airspeed, you end up paying a price. In this case it is the airplane's superior stalling characteristics; they become tricky, abrupt, and unpredictable in even mildly turbulent air. Therefore, except in conditions of dense, calm air, the wise pilot hesitates to slow the STOL-Cub as much as he might a stock one, which no doubt accounts for the fact that I—along with most other high-time Cub pilots—have yet to meet a pilot who can operate one of the wing-modified Cubs out of strips any shorter than those we use with our standard Cubs. In fact, the reverse is usually the case. You can't really *buy* short-field capabilities, no matter what the ads say: you have to earn them through laboriously acquired skills.

Now to the numbers, which bush pilots do not take too seriously except for the best rate and angle of climb speeds:

Takeoff Run (ft) (full flaps)	200
Takeoff Run over 50 ft obstacle (ft) (full flaps)	500
Best Rate of Climb Speed (mph)	75
Rate of Climb (ft per min) (no flaps)	960
Best Angle of Climb Speed (mph) (full flaps)	45
Best Angle of Climb (ratio)	1 to 5
Service Ceiling (ft)	19,000
Absolute Ceiling (ft)	21,300
Top Speed (mph)	130
Cruising Speed (75% power; mph)	115
Cruising Range (75% power)	460
Fuel Consumption (75% power; gph)	9
Stalling Speed (mph) (flaps)	43
Landing Roll (ft)	350
Gross Weight (lbs)	1,750
Empty Weight (standard)	998
Useful Load (lbs)	752

Note that the Super Cub, like the Cessna 180/185, is equipped with mechanical flaps operated with a Johnson Bar. This feature is of special interest to the pro working the shortest strips, and its use will be covered in Section II.

Chapter 9

The Cessna 180/185

Along with the Super Cub and the Cessna 206, the Cessnas 180 and 185 form the backbone of the professional bush pilot's equipment list in Alaska. However, while various examples in this book refer to the Super Cub or the Cessna 206 specifically, the Cessnas 180 and 185 (unless otherwise noted) will be discussed as if they were the same airplane. As a form of shorthand, I will call them the Cessna 180/185.

This is more than mere convenience, for just as Detroit offers the same automobile or truck with the option of different size engines, so Cessna offered the same basic airframe with the choice of small or large engines.

The only relevant difference between the Cessnas 180 and 185 is engine power. Otherwise, they share the same fuselage, wings, control surfaces, horizontal stabilizer, and main landing gear (the Cessna 185 has a locking tailwheel). The late models of both are six-place high-wing conventional-gear airplanes of metal construction.

The *Cessna 180* is equipped with a carbureted Continental 0-470-U engine rated at 230 horsepower.

The *Cessna 185* is equipped with a fuel-injected Continental IO-520-D engine rated at 285 horsepower continuous, with 300 allowable for five minutes at takeoff.

A look at the specifications reveals what the different engines have to offer. These specifications are at gross weight; with the Cessna 180 and 185 carrying the same load, the takeoff, climb, and ceiling would be different, with all advantage to the Cessna 185.

The Cessna 180

Takeoff Run (ft)	625
Takeoff Run over 50 ft Barrier (ft)	1,205
Rate of Climb (ft per min)	1,100
Service Ceiling (ft)	17,700
Top Speed (knots)	148
Cruising Speed (75% power; knots)	141
Cruising Range (75% power; nautical)	725
Cruising Range (maximum; nautical)	890
Stalling Speed (knots)	48
Landing Roll (ft)	480
Gross Weight (lbs)	2,800
Empty Weight (standard)	1,648
Useful Load (lbs)	1,152

The Cessna 185

Takeoff Run (ft)	770
Takeoff Run over 50 ft Barrier (ft)	1,365
Rate of Climb (ft per min)	1,010
Service Ceiling (ft)	17,150
Top Speed (knots)	155
Cruising Speed (75% power; knots)	145
Cruising Range (75% power; nautical)	585
Cruising Range (maximum; nautical)	720
Stalling Speed (knots)	49
Landing Roll (ft)	480
Gross Weight (lbs)	3,350
Empty Weight (standard)	1,687
Useful Load (lbs)	1,663

As the figures indicate, each of these airplanes fits a slightly different niche in the overall fabric of the trade and is used for slightly different purposes. Generally, this is how they stack up:

The Cessna 180. Because the Cessna 180 (see illustrations 16 and 17) has a smaller engine than the Cessna 185, it also has somewhat better fuel-burn figures. This gives it better range (890 nm versus 720 nm), and you can take advantage of this under the following conditions:

* On those missions that require you to fly into very remote areas where refueling facilities are not available and you are unable, for one reason or another, to land en route and refuel from five-gallon cans carried in the cabin. For example, most of the Canadian Northwest Territories falls into this category when you have to operate on wheels.

* Under low instrument meteorological conditions in most areas of Alaska and Northern Canada. In these areas, IFR alternates are generally few-and-far-between, and maximum range capabilities all too often spell the difference between go and no-go. (An auxiliary fuel tank such as Flint Aero offers for the Cessna 180/185, adding 23 usable gallons, could come in handy here.)

On economical grounds, the Cessna 180 gets the nod when the airplane is on wheels and based in a local area where trip lengths rarely exceed 200 miles. In this situation, gross weight will be limited primarily by cabin volume and fuel weight will seldom go over 182 pounds. As the Cessna 180 is about $15 per hour cheaper to operate than the Cessna 185 (because of fuel and engine costs), you can bid jobs at more attractive rates than you could if you had to use a Cessna 185 for this purpose. In difficult economic times (such as Alaska is going through as this book is being written), this price break, small as it is, might offer some assistance to your customers.

The Cessna 185 The Cessna 185 provides power over economy and range. When missions require you to operate on wheel-skis or floats, it's difficult to justify using a Cessna 180 if a Cessna 185 is available. Basically, what you get is:

* Markedly better takeoff and load-carrying performance on wheel-skis and floats. Not only do wheel-skis and floats seriously compromise your payload because of their weight, they also require large amounts of power

to overcome surface drag during the takeoff run. Here, there is no comparison between the Cessna 180 and 185; the Cessna 185 is always the winner.

* The ability to carry full cabin loads for reasonable distances even with the weight penalties imposed through the use of floats and wheel-skis.

With fairly light loads—about what would be considered gross for a Super Cub—the Cessna 180/185 is capable of operating out of gravel bars measuring 500 feet in length under Northern conditions (see Section II, and Chapter 25 in Section III).

Tires. The best tire for the Cessna 180/185 is the 8:50x6. The 25-inch tundra tires should be avoided as, when inflated correctly, they are too flexible for the Cessna spring gear. This is especially true when used under conditions of strong crosswinds on paved strips.

Tailwheels. Some operators use the 5:00x5 tire on an oversized fork with the Cleveland wheel. This is an improvement, especially on soft rocky strips where the standard tire assembly not only has a tendency to dig into the ground like a plow, but gets rather chipped up.

Oil quick-drain door. One modification I especially like is to install a small door, similar to the one used to add oil, on the lower right side of the cowling where the oil quick-drain can easily be reached. This way, during winter when temperatures get well below freezing and you need to drain your oil into a clean can to store overnight by the stove in your cabin (to keep it warm for the next morning's flight), you can do it quickly, easily, and without mess. (See Chapter 12.)

Beyond this, no other modifications are needed.

Chapter 10

The Cessna 206

All pilots have their favorite airplanes; mine are the PA-18-150 Super Cub and the turbocharged Cessna TU206G. I prefer the Super Cub for the low and the slow, the contour flying and serious off-airport landings; I like the Cessna 206 for carrying loads, traveling distances, and flying on instruments.

However, there is a small difference of opinion regarding the Cessna 206, one that has considerable validity on both sides: which model is best for the bush operator—the turbo or the non-turbo? As with all questions of this nature, the answer is: it depends.

The non-turbo Cessna 206. Ordinary air taxi work—whether on wheels, skis, or floats—involves intense competitive pressures, and these pressures force you to make your aircraft and equipment selections with your eyes focused firmly on the bottom line. Acquisition cost, fuel burn, maintenance load, TBO, and avionics all come under close scrutiny. You must also consider simplicity of operation, for experienced and knowledgeable pilots are expensive to hire and hard to keep, and an engine that requires a delicate and knowing touch to operate properly (the penalty for aggressive throttle use, over-leaning, shock-cooling, or ham-handed over-boosting may well involve a visit to the loan department of your bank) is not a wise business investment. Let's take a look at the specs, for both models, and see how they stack up on paper:

The Cessna 206

Takeoff Run (ft)	900
Takeoff Run over 50 ft Barrier (ft)	1,780
Rate of Climb (fpm)	920
Service Ceiling, wheels (ft)	14,800
Service Ceiling, floats (ft)	13,900
Top Speed (knots)	155
Cruising Speed (75% power; knots)	147
Cruising Range, wheels (75% power; nautical)	680
Cruising Range, floats (75% power; nautical)	615
Cruising Range, wheels (maximum; nautical)	900
Cruising Range, floats (maximum; nautical)	770
Stalling Speed, wheels, (knots)	54
Stalling Speed, floats, (knots)	52
Landing Roll (ft)	735
Gross Weight (lbs)	3,600
Empty Weight (standard)	1,882
Useful Load (lbs)	1,718
Engine TBO (hrs)	1,700

The Turbo Cessna 206

Takeoff Run (ft)	835
Takeoff Run over 50 ft Barrier (ft)	1,640
Rate of Climb (ft per min)	1,010
Service Ceiling, wheels (ft)	27,000
Service Ceiling, floats (ft)	25,600
Top Speed (knots)	174
Cruising Speed (80% power; knots)	167
Cruising Range, wheels (80% power; nautical)	640
Cruising Range, floats (80% power; nautical)	550

Cruising Range, wheels (maximum; nautical)	805
Cruising Range, floats (maximum; nautical)	690
Stalling Speed, wheels (knots)	54
Stalling Speed, floats (knots)	52
Landing Roll (ft)	735
Gross Weight (lbs)	3,600
Empty Weight (standard)	2,000
Useful Load (lbs)	1,600
Engine TBO (hrs)	1,400

As the numbers show, the basic differences are service ceiling, range, and TBO. Therefore, for the VFR-only operator who does not require high-altitude capability, the standard Cessna 206 is the best choice. Inexpensive, simple, and easy to fly for even low-time commercial pilots, it is a money-making machine *par excellence* for the air taxi operator.

The Cessna Turbo 206. With the Cessna Turbo 206 you move into a different league entirely. Here is a machine that can operate comfortably out of a 900 foot gravel bar if not too heavily loaded, climb rapidly into the flight levels to top weather, then make an instrument approach to RVR 1800. But all this capability costs, and so you have higher fuel burn, an expensive, delicate engine that must be handled with kid gloves, and a TBO of 1400 hours which only the best operators reach. To that, if you expect to get full value out of its IFR capabilities, must be added the cost of sophisticated avionics, and a pilot who knows how to operate them. Obviously not the type of equipment seen on the usual bush ramp or gravel bar, it is reserved for the small specialty operator who has a quality-conscious clientele that demands the best in pilots and equipment and is willing to pay the premium necessary to obtain them. Here, you are operating the way a top corporate flight department would, where the emphasis is on passenger safety, schedule reliability, and quality equipment, not the bottom line. (See illustrations 24 and 25)

Instrumentation (IFR). The Cessna 206 is a great instrument platform, stable and sure, and when properly equipped it can handle many weather problems with ease. My 1985 Turbo 206, used for serious instrument flying, is currently outfitted as follows (see illustration 26):

* Dual Nav/Comms with VOR/LOC/GS reception (King KX 155)
* DME (King KN 64)
* Loran (King KLN 88)
* Flight Control System (King KFC-150)
* Altitude Preselect (King KAS 297B)
* Transponder/Encoding Altimeter (King KT 76A/KEA 130A)
* ADF (King KR 87)
* HSI (King KCS 55A)
* RMI (King KI 229)
* WX-1000+ Weather Mapping System (BFG/Foster)
* CFS 1000A Fuel Management Computer (SDI/Hoskins)
* Graphic Engine Monitor (6 cylinder EGT, CHT; TIT) (Insight GEM)
* Digital Density Altitude, Outside Air Temperature (F & C), Pressure Altitude, and Aircraft Voltage Gauge (Davtron)
* AltAlert altitude alert and fuel-burn timer (Icarus)
* Electronic Carbon Monoxide Detector (Paragon)
* Bose headsets (2) (front seats)
* David Clark headsets (2) (rear seats)
* ISOCOM intercomm (David Clark)

As you can see, I believe in just the bare necessities…

Tires (main gear). As with the Cessna 180/185, the 8:50x6 6-ply tires are best for the main gear. If a proposed landing area is too soft or rough for them, use a Cub or Cessna 180/185 instead; it is no place to take a Cessna 206.

Tires (nose gear). Here, there's a choice of three tire sizes. The 6:00x6 tire is a factory option with the Cessna oversized nose-gear fork, and many operators find it to be adequate. However, since this fork will also accommodate the larger 7:00x6 tire, those who operate regularly out of villages or camps where strips are soft, rough, and poorly maintained often specify it instead.

If your job demands that you operate on beaches or other soft, sandy areas, it sometimes helps to go to the 8:00x6 tire. This will require a special, modified Piper nose fork; Airglas Engineering, of Anchorage, Alaska, converts forks for this purpose. (The disadvantage here, of course, is the way the 8:00x6 tire handles on paved runways.)

Long-range fuel tanks. For certain trips into the more interesting areas of Canada's Northwest Territories, where fuel is difficult to come by and the chance for unforecast IMC high, long-range tanks are a necessity. These tanks are also required for many instrument flights, where viable alternates often demand truly creative—some would say heroic—flight plans. Fortunately, the Cessna 206 has a nice choice of long-range tanks available. Here are two (of quite different capacity and design philosophy) that work well:

* Flint Aero. The old boy on the block, these wingtip tanks have been around for years. Depending on the model Cessna 206 you have, they add either 27, or 29.8, usable gallons, and, in the process, an additional three feet to your wingspan. Also included, for the F and G models, is a STC for a 200 lb increase in gross weight.

This system requires two electric transfer fuel pumps, each with its own on/off switch and fuel gauge, and the pilot must manually transfer fuel from the tip tanks to the main tanks when needed.

I have these tanks installed on my '85 Turbo 206, and like the extended range I get with them, as well as the additional stability the tip tanks provide in turbulence when they are full. However, three factors should be noted:

1) The tanks are quite difficult to install properly in the time frame Flint claims (it takes most good shops that are interested in quality work about 80 hours).

2) The caps have a poorly-designed vent hole on the top that leaks fuel when taxiing, making turns, or flying in rough air; fuel loss stops at about .6 gallon per tank, so this must be taken into consideration during flight planning. Flint has promised to redesign these caps, but it has been a number of years now since he said he would do this, and, as with the check in the mail, I am still waiting.

3) The added wing span provides a slight increase in STOL capabilities, at the expense of making some hangars unusable for the airplane. Also, of course, after installing the tanks, the airplane's wing covers must be modified because of the increased span.

* Uvalde. Manufactured to Part 23 standards, these internal bladder tanks—there are three for each wing, and they fit just outboard of the main tanks—add 54 gallons to your fuel capacity. A passive system, there are no

switches to flip or fuel management techniques to learn. Unfortunately, the Uvalde tanks do not come with a gross weight increase, so the weight of the additional fuel (324 lbs) must come directly out of your payload.

Engine analyzers and fuel computers. For all serious, maximum-range operations, I recommend that an engine analyzer, such as the Insight GEM, *and* a fuel computer, such as the SDI/Hoskins or Shadin Miniflo, be used. Only in this way can you manage your fuel burn, and keep up to date on the progress of your flight, with the precision required.

Deicing/anti-icing equipment. You do your best to stay out of icing conditions, but sometimes, no matter what the forecasts or PIREPS had to say, your best efforts fail and ice finds you anyway. Therefore, deicing/anti-icing equipment is high on the priority list, even though at this time the FAA will not certify the Cessna 206 for flight into *known* icing conditions. However, you do have a bit of help available for those times when icing conditions are mild—stratus with low tops; a layer of steam fog, etc.—and your accidental exposure brief:

* Propeller. Most experts agree that if you had to make a choice between wing or prop deicing/anti-icing equipment, prop it would be, for with the prop clean, as Bob Buck observes in his fine book *Weather Flying,* "you can pull a lot of ice-covered airplane around the sky."

Since a propeller anti-ice system is a factory option, serious instrument pilots always make a point of having one installed on their airplane.

* Windshield. Anti-icing for the windshield is useful when you fly from a deep bush location that is "severe clear" but very cold into a coastal area where an approach has to be made to minimums through a layer of steam fog. This steam fog will just frost the windshield but that is enough, and while the airplane's defroster (if placed on high 20 minutes or so prior to the approach) is usually adequate to keep at least a small area directly above the heat vent clear, sometimes it isn't. On such occasions, the alternatives are to land by looking through an opened side window, or to use a windshield anti-icing system. The latter is far more elegant, and since a detachable electrically-heated glass panel is a factory option, it should be given serious consideration if you need this type of equipment.

* Combined wing, tail, struts, prop, and windshield anti-icing. Kohlman Aviation offers the British TKS ice protection system for the Cessna 206, which uses a glycol-based fluid to prevent or remove ice buildup. This well-proven system, whose lineage can be traced back to the mid-1930s

(currently it is being used on airplanes as diverse as the BAe 125 and Short Skyvan), is certified on several light airplanes for *known-icing conditions* in Europe. Unfortunately, the FAA, in its infinite wisdom, has decreed that additional backups, primarily a second electrical system, be required if you wish United States certification for your "small Cessnas, Mooneys, and Bonanzas."

So, though it is a proven system that is highly effective and can be installed under a STC, it is not at the present time legal for known icing in this country. Nonetheless, since real-time icing reports are rare and hard to come by in the remote north, and icing by its very nature is difficult to forecast accurately, the protection that the TKS system offers can be of value to some bush operators.

Dual Vacuum Systems. Factory options, they are a must for IFR.

Cargo Tiedowns. These tiedown blocks—they bolt to the seat rails—are an important Cessna option for aircraft used in cargo operations. I generally use 12 of them, and *always* secure my cargo well—using a custom-made heavy-duty cargo-restraint system manufactured out of 2" nylon webbing by Cargo Systems Company of Dallas, Texas—even though it adds a little extra time to the loading and unloading process.

Section I, Part IV

Special Operating Procedures

Special Operating Procedures

Whether your base is Alaska or Minnesota, Inuvik or Chicago, winter brings problems with frost, snow, and ice contaminating your airframe; and the cold requires that you not only preheat your engine, but also utilize special operating procedures while in flight.

In addition to these problems, the Northern pilot must deal with the difficulties created by the vastness of the Arctic, its small population and widely scattered communities. Here, far more than elsewhere, you must take into consideration range and alternates, both VFR and IFR, and include them not only in your basic operating procedures but within your mindset as well. For bush flying is, if nothing else, a thinking man's game, and how well you think (and look ahead, anticipating potential problems) determines not only how successful you are in completing your assigned flights safely and economically, but also your very survival.

In bush flying, it is the small (what some call "minor") details that most often bite you. That slight dusting of frost left on the wings combined with a short runway, for example, or the unexpected weather that suddenly leaves you trapped without an "out," a safe haven where you can take refuge.

So, let's take a look at some of the more prosaic details of the bush pilot's profession, those without glamour or mystique, only utility.

Chapter 11

Problems with Frost, Snow, and Ice

There is a mystery, a strangeness, about it. Every winter the same scenario repeats itself over and over again without fail or the slightest respite. An airplane with frost, snow, or ice contaminating its wing and tail surfaces taxis to the runway and moves into position. Power is applied, the plane lifts lumberingly into the air, staggers through ground-effect, stalls, and spins in. Or, alternatively, mushes into the trees off the end of the runway. Gets "rolled in a ball" as they say in the business.

This scenario is not in the least confined to light airplanes and amateur pilots: in 1987, 28 people were killed and another 28 seriously injured at Denver after a crash of a Continental Airlines DC-9-14 because of snow and ice on the wings. And two years later at Dryden, Air Ontario did no better, with a death toll of 24 of their 69 passengers.

Witnesses reported "a considerable amount of snow and ice" on the wings of Air Ontario's F-28 during the takeoff roll; the pilots were offered deicing services prior to takeoff but "declined." As for the Continental pilots, they apparently sat unconcerned in a nice warm cabin in a wet snowfall (classified as "moderate") at the runway threshold for 27 minutes after having their plane deiced, even though the DC-9-14 is notorious for its inability to withstand even $1/32$ of an inch of wing contamination, and Company policy demands *at least* a visual check of the wings prior to takeoff if the delay has been longer than 20 minutes.

What is going on? What could be preventing these pilots from responding to this situation, especially when the aerodynamics of the situation are so clear, so terribly unambiguous?

If anyone who reads this has the answer, I would like to learn what it is, though I have a sneaky hunch this is a problem for the boys (and girls) who specialize in abnormal psychology. Surely it can't just be a product of plain (plane?) stupidity.

As for those who know better, their problem is simply deciding which of the many techniques available is the best to use under the circumstances.

Prevention

Wing covers

Wing covers, in the bush or on the ramp, are your major line of defense, and, like many simple and obvious remedies, they work surprisingly well.

These covers, which are now made of nylon—in ye olden times, they were made of stiff, heavy canvas and were quite bulky—are designed to slip over the top surface of the wings and fasten underneath with either bungee cords, nylon straps, or ropes (see illustrations 18 and 27).

The usual procedure is to place them on the airplane after the last flight of the day, while tying it down. Then, on the following day, just prior to flight, they are removed. When not on the airplane, they are stored in a small stuff-bag which is left at home base, unless it is anticipated that the airplane will be away for the night.

Like anything else, though, wing covers are not a perfect solution to all problems and here the main one is a snowstorm that turns to slush or rain and then freezes. This type of situation usually produces the following problems:

1) Soaked wing covers that are now frozen to the wings.

2) A thick layer of rime ice on the top surface of the wing covers, tail, and the rest of the airplane's structure.

3) Accumulations of ice on the bottom rear edges of the control surfaces caused by water building up and then freezing on the underside of the wing covers where the straps attach. This ice, if allowed to remain, can cause control flutter which is very dangerous; it must be removed prior to flight.

Out in the remote bush, far from a heated hangar, these problems are not nearly as insurmountable as they appear. A little methodical effort will soon put things right once again:

* Wings. The wing covers will peel off the wings leaving behind a thin coating of ice which will usually be of little serious consequence since it generally follows closely the curvature of the wings, therefore disrupting the airflow only in subtle ways. Once in the air you can seek out a warm layer above the inversion, and, if found, melt off this coating.

* Bottom rear edges of the control surfaces. Here you will have to remove it by hand, usually not all that much of a job.

* Horizontal tail. Here, too, you will have to manually remove the contamination, and this is done by rubbing it loose with your gloved hands, or a piece of folded rope. Crude, but it works, and if done carefully it will not damage the paint. With fabric airplanes the flexing of the fabric will easily break the ice loose; with metal airplanes it is slightly more of a job.

* Prop and spinner. The propeller, an airfoil, will have to be carefully cleaned. The easiest method is to take a piece of wood like an old axe or broom handle and carefully rub it up and down the blades until all ice has been removed.

The spinner must also be checked, for often melting snow can collect in it creating an out-of-balance condition. It helps to align the prop horizontally prior to the storm—or, if leaving it partially-vertical, cover it with a plastic garbage bag—as the main route of this moisture is down the blade when the prop is left in its natural position after shutting down the engine.

* Windshield. This works the same way as the wings. Peeling off the cover (described below) leaves a thin layer of ice. Usually this takes the form of streaks and patches, leaving a spot or two where the visibility is good enough to allow takeoff. Once in the air, a few minutes of defroster action will clean it.

* Hinges, moving parts, and the pitot/static system. The rest of your airplane must be gone over carefully to make sure there is no ice that could

interfere with the proper operation of any system or control. If found, it will have to be removed by hand.

* Wing covers. The final step is to clean these covers; this you will do also for the engine, windshield, and tail covers, which will be described next. The best way to do this is to remove as much snow and ice as possible outside, and then hang them in a warm building to thaw and dry. City dwellers can just dump them in a clothes dryer, turn it on, and go about their business; when they get back after their flight, their covers will be nice and dry, waiting to be used once again.

High winds, as well, cause problems with wing covers that are not properly installed. Bungee cords and metal hooks used for convenience to attach the covers loosen when temperature variations and age manage to cause them to stretch and become inelastic. Lost wing covers, damaged paint and fabric, and, sometimes, even broken windows—from the hooks banging around during a storm—result.

The best way to forestall these problems is to throw away the bungee cords and hooks and replace them with ¼ inch nylon cords, or one-inch nylon straps—the soft type you can tie, like those Cessna uses with its cargo nets. In this way, you will be able to carefully adjust the tension of each cord to its optimum setting without hooks or buckles, and the wing covers will always fit properly.

In my experience, wing covers fastened this way will stay in place in winds officially clocked in excess of 80 knots. As an added advantage for the Super Cub, they will act as control locks for the flaps and ailerons.

It takes about ten minutes to install a set of wing covers properly. Like tying your airplane down correctly so it will ride out a serious storm, the regular use of wing covers is well worth while.

Tail covers

Tail covers function like wing covers but are rarely used in the bush because of the time and effort involved in putting them on and taking them off. I use them in freezing rain, but rarely otherwise.

Windshield covers

Windshield covers are used just like wing covers, are made of the same material, and serve the same purpose. However, during windstorms, it's better not to use them—they can scratch the windshield if allowed to rub against it.

As with wing covers, replacement of the bungee cords and hooks with nylon cords or straps is advised.

Engine covers

Engine covers are made of a nylon shell padded on the inside with a synthetic insulation. They are primarily used to retain heat in the engine compartment during short ground waits between flights, and while preheating the engine (see the next chapter).

A secondary use is to keep snow and ice off the cowling and out of the engine compartment. Like the wing covers, the engine cover does a good job of keeping a surface clean, but once the wind gets above 60 knots or so, blowing snow will work its way through the overlapping edges and some engine compartment contamination will occur. Usually, preheating will resolve this problem without difficulty, if you are using a combustion method, which frequently is the case in remote areas.

Hangars

Of course, all these considerations are moot if you have the luxury of a hangar and the airplane is on wheels or wheel-skis—although heated hangars have their own problems, described below. But there are few hangars in the bush, and many airplanes on board-skis. So you do the best you can with what you have, and I must confess that in over 22 years of flying Alaska and Northern Canada the only time any of my airplanes ever saw the inside of a hangar was when they were in Anchorage for maintenance.

Cure

Based far from the support of civilization, bush pilots have a tendency to be far more careful and methodical than city pilots in the way they secure their airplanes after the last flight of the day. As a result, they rarely have to contend with contamination problems—except for the previously mentioned freezing rain/snow situation. However, for pilots visiting town who get caught in an unexpected snow storm, or who live in town but somehow neglect to put on their wing covers, the following techniques can prove useful.

Hangars

If your snow or ice-contamination problem occurs at an airport where there are maintenance shops available, a bad case is easily—but expensively—solved by shoving the airplane into a heated hangar so it can thaw out.

However, there is one major caveat concerning the use of heated hangars to defrost cold-soaked ice and snow-covered airplanes: they must be left in the hangar not only long enough for all the snow and ice to melt off, but also *long enough to get completely dry both inside and out.* If this is not done, you risk exposing yourself to the following problems:

* Control-flutter caused by refreezing melted snow and ice that has collected but not completely drained from the elevator, ailerons, etc.

* Control binding caused by the freezing of wet pulleys, etc.

* Frozen condensation in the fuel thawing, and then refreezing, so as to block fuel flow. The tanks and all sumps should be well drained each time the airplane spends more than a couple of hours in above-freezing temperatures during the winter. When temperatures are below freezing, it is generally a poor idea to drain the sumps because they will often fail to reseal properly leaving one with a fuel leak.

* With the Super Cub, snow is often blown into the fuselage through the stabilizer adjustment slot, and, in serious cases, this can form a block of drifted snow stretching from the battery-box to the bellcranks and upper control cables. This snow must be removed before taking the airplane from the hangar. If it is allowed to partially thaw and then refreeze, it could not only interfere with control movements but also place a serious weight at the very aft-end of the c.g. arm.

A heated hangar is nice, but it must be used carefully and intelligently during cold weather.

Hot water

Using hot water to defrost an airplane can get one in hot water. I will sometimes use it on the windshield if conditions are bad, but *nowhere* else. The reason should be obvious.

Broom

When the snow is dry, or wet with temperatures above freezing, using a broom to remove the snow works very well. In the bush, even with wing covers installed, it is standard practice to brush off all the snow you can prior to removing them. A good broom—with soft bristles that will not damage the airplane's paint—can be considered part of the winter equipment list for all remote-area bush pilots.

Rope

The following technique, while approaching the Mickey Mouse level, can be used to smooth down heavy frost so the airplane will fly. It should not be used except in emergencies, when you get caught by an unexpected frost in some remote area and there are no other options available.

What you do is take one of your tiedown ropes, pass it over a wing, and then work it back and forth using its abrasive characteristics to wear down the frost. It is more a polishing job than a removal job, and the most important part of the wing is the leading edge. Well polished—on the airplanes used for bush flying—the wing curvature will be correct enough to allow for flight, and a few hours in the air will finish the job. A word of advice: keep the c.g. in the mid-range and the load light until the wings are completely clean, and do not try to horse the airplane off the ground early on takeoff.

Chemicals

These have no place in the bush pilot's world, and I have never seen or heard of anyone using glycol or automotive deicer. The cost of purchasing it, transporting it into the bush, and using it on a daily basis makes as much sense as hauling out an electric generator to operate an electric carving knife or can opener. This stuff is best left to the airlines, with their trucks, spraying devices, and teams of workers.

Chapter 12

Preheating Equipment and Methods

Engine Compartment

Aviation fuel is distilled with a much lower vapor pressure than automobile fuel so that during the climb to altitude engine failure will not be experienced due to vapor lock. What this means from an operational point of view is that once temperatures get down to about 15° F/-9° C, the engine will require preheating so the fuel can properly vaporize for ignition.

Compounding the problem, at about -5° F/-21° C most oils (the exceptions being synthetics like Mobil AV-1) become too thick to lubricate the cylinders and bearings properly, and if you could manage to start your engine at this temperature (by heavy priming, or a bit of mild preheating), the wear that would take place until the oil began to flow would probably be equivalent to more than 50 hours of normal operation.

(Mike Busch, the Cessna Pilots Association's engine specialist, claims that one cold start at the lower temperature extremes can cause as much wear as 500 hours of normal cruise operation. This takes into consideration the metal-to-metal contact between the piston and cylinder walls caused by the uneven expansion of the aluminum pistons and steel cylinders as the engine goes rapidly from very cold to very hot.)

If you have a turbocharged engine you are well advised, even when using multi-grade oils, to preheat below 30° F/-1° C: the turbocharger is very oil-critical. As for non-turbocharged equipment, preheating is mandatory for Continental and Lycoming piston engines below 10° F/-12° C. I preheat all my engines, turbocharged or not, below 30° F/-1° C: I like to get my full TBO.

There are no convenient ways around these problems, and, like the FAA and taxes, they are a fact of aviation life. In this chapter we will take a look at the best methods available to both the remote-area bush pilot and the city-based airport pilot when operating under daily conditions that require preheating. However, before doing so, let's continue our examination of a piece of equipment (mentioned in the previous chapter) that is necessary to both areas: the insulated engine cover.

Engine covers

Now made out of nylon like our wing and windshield covers, engine covers are padded with about an inch of synthetic insulation and are designed to fit snugly over the engine cowling, fastening with Velcro strips, nylon ties, and bungee cords (see illustrations 28 and 29). Used properly, they also serve a number of related uses during the cold, stormy months of winter.

* During periods of deep cold (-20° F/-29° C and below), they keep the engine warm between flights, so preheating does not have to be repeated throughout the day. If placed on the cowling as soon as the engine is shut down, oil temperatures will remain in the green for about three to six hours, depending on whether you are using a thin cover like the Tanair, or a thick cover like those manufactured by Alaska Wing Covers. In those rare emergencies where you are forced to RON someplace where preheating facilities are not available and the portable preheating equipment that should be in the airplane's survival kit has been left behind, the running of the engine for a few minutes every few hours will keep it warm enough to remain operational. The only problem you need worry about here—aside

from lack of sleep—is the chance of running low on fuel if emergency conditions, usually weather, continue too long.

* At urban bases or public airports where electricity is available, the covers are used in combination with an electric heater to maintain engine heat for 'round-the-clock availability.

* At base camp, in the bush, or on airports without available electricity, engine covers are used in combination with combustion heaters to focus and maintain heat flow (see illustration 28).

* And, as mentioned in the previous chapter, they are used to keep the engine compartment free of blowing snow during wind storms.

Electric heaters

In "ye olden days" bush pilots often used little 750 watt "car warmer" heaters that cost about $20. These heaters typically measured about 4"x6" x7" and contained a metal heating element, a temperature-controlled safety switch, and a small fan. One or two placed in the engine compartment after the last flight of the day, with the cowling then covered by an engine cover, did a fine job, and they were the standard until the fine Tanis preheat system came on the market. Now the car warmers are mostly relegated to the cockpit, where they help keep it and the instruments warm.

The Tanis preheating system is a light-weight (one to two pounds, depending on application) system of thermal pads that attach to the crank-case and oil pan, along with heater elements that attach to the cylinders. In the better installations, a flush-mount power receptacle is installed in a non-structural skin area (such as an engine baffle) on airplanes like the Super Cub, Cherokee Six, or Bonanza that have cowlings that open easily for pre-flight inspections (see illustration 30). On airplanes like the C-180/185/206 where you have to use a screwdriver to get the cowling open, a flush plug door, similar to the ground service plug receptacle for an external power source, is installed. The standard Tanis TAS-100 system is good down to about -30° F/-34° C; the Tanis "Super System," which I use, is good down to -65° F/-54° C. Recommended highly.

An emergency technique

Sometimes a pilot will get caught away from home base without his combustion preheating equipment, but will have electricity available. Here the thing to do is borrow a droplight (using a 100 or 150 watt bulb) from the local maintenance shop and place it in the engine compartment. With the engine cover in place, this will keep the engine warm indefinitely as long as the temperature does not drop too far.

Combustion heaters (in the city)

Combustion heaters range from the large Herman Nelson and space heaters used by FBOs for transient aircraft (for prices ranging from $25.00 to $40.00 per engine) to the small portable units used by the bush pilots. All these units need to be used wisely, for applying just enough heat to get an engine started once it has been cold-soaked overnight is a fine way to damage it, and bring on an early overhaul.

* Herman Nelson. These large and efficient heaters, usually seen on wheeled carts, were originally designed for DC-3 size aircraft and are used extensively all over the North by FBOs and charter outfits for the preheating of transient aircraft, company Twin Otters, Beavers, etc. They are combustion heaters which use a forced-air fan operated by either an electric or gasoline motor to drive heated air through long ducts into the engine compartment, battery compartment, and cabin of the airplane.

When used on light aircraft under conditions of deep cold (-20° F/-29° C and below), these heaters will easily get your engine warm enough to start before the oil has an opportunity to completely thaw within the crankcase. This can present serious problems if, when you are in the air, a mass of half-frozen oil blocks the oil pump. In fact, as the old saying goes, it could ruin your whole day.

If circumstances do not allow you to use the Herman Nelson longer than the bare minimum required to start the engine, the following emergency procedure should be followed:

1) Start the engine and, with a careful eye on the oil pressure gauge, warm it up until the cylinder head temperatures reach about ¾ into the green.

2) Shut the engine down and cover it with the engine cover.

3) Wait about 40 minutes until the engine's heat can transfer to the oil, thawing it.

One final point. When an airplane engine has been preheated, the oil in the dipstick area seems to thaw first, so it is a mistake to check the oil, note that it runs free on the dipstick, and therefore assume that it has been adequately thawed. If you doubt this, a little test will demonstrate the principle involved. Take a can of oil that has been left outside in -40° F/-40° C temperatures and bring it in to where the wood or oil stove is. Punch a hole in the top of the can and place it on the hot stove. Let it sit there until the outside of the can is too hot to touch and a nail dipped inside the can shows free-flowing oil. Then pour the oil from the can into another can. Depending on timing, there will be a large or small lump of still-frozen oil in the can; this is a sample of what the oil inside the engine is like.

* Space heaters. These electric/combustion heaters, refugees from the construction industry, are also used by FBOs and charter outfits, mostly in Alaska, for company and transient aircraft. They are placed on the ground under the engine compartment, which is then draped with a blanket or piece of canvas.

While these heaters are hardly in the Herman Nelson class, they do a good job when suddenly called upon to heat an engine, and all you need remember is the caveat concerning the frozen oil.

Combustion heaters (in the bush)

Preheating an engine in a wilderness environment demands that your combustion heater be dependable, durable, and portable.

For dependability, avoid anything too fancy or complicated, including those cute (and expensive) little blower heaters operated by a 12 or 24 volt DC motor that are designed to be hooked up to the airplane's electrical system. In conditions of deep cold, these will often not work due to a cold-soaked battery.

For durability, avoid most of the camp stoves now on the market. These are designed for the weekend camper; they usually lack quality construction and are not something you would wish to stake your life on.

For portability, choose a heat source that is small and light enough to be carried in the airplane's winter survival kit (see Chapter 24). This is important, for the professional bush pilot will every now and then find himself stuck overnight someplace due to a fast-moving weather system and will require the services of his engine preheater once the front moves through.

The best bush preheating rig is one that uses a simple old-fashioned gasoline blowtorch that can be operated on avgas taken directly from the airplane's fuel sumps. Since this is a systems approach, if you wish to use a different heat source you will find little difficulty in adapting this technique to your purpose. For the convenience of those who are unable to locate gasoline-powered blowtorches—they are hard to find these modern days—several alternative heat sources have been listed, though they all fall far below the level of utility that the blowtorches achieve.

* Blowtorch and stovepipe method. The equipment list for this system consists of a small fire extinguisher, one or two gasoline blowtorches, three lengths of 3" stovepipe, an insulated engine cover, and a two-or five-gallon metal gasoline can. Here is how the system works:

1) After the last flight of the day, the oil is drained from the engine into the gasoline can. Quick-drains should be installed on all airplanes in the fleet to facilitate this daily chore.

2) The can of oil is taken inside your office or home, where it is placed near the stove so it can be kept warm until needed again.

3) When the airplane is to be called into service, the can of oil, with the lid removed, is placed on top of the stove and heated to about 122° F/50° C.

4) While the oil is warming, take the three lengths of 3" stovepipe and fasten them together. Place one end inside the engine compartment and temporarily rest the other end on the ground. The best place for the pipe to go into the Super Cub's engine compartment is on the right side by the brace to the exhaust stack. With the Cessna 180/185, it is best to drop the right cowl-flap and place the pipe there (see illustrations 28 and 31).

5) Carefully light the blowtorch away from the airplane, and when it is operating properly place the nozzle into the lower end of the stovepipe.

With the Cessna 180/185 and 206 you should use two blowtorches, and these *must* be placed in the same pipe (see illustration 31). Never use two pipes, one going to the right side cowl-flap and the other going to the left side cowl-flap, because of the danger that one of the blowtorches might go out, while still spraying fuel into the stovepipe. The second torch could then ignite the fuel inside the engine compartment. With both torches in the same stovepipe this danger is averted.

6) Adjust the insulated engine cover over the top and sides of the engine compartment so the right cooling air-inlet, behind the prop, is covered and the left air-inlet is open (see illustration 28). This is done to allow the heated air to pass over the engine and out the far side, taking the cold with it.

Remember, heat circulation is the name of the game. When using combustion heaters, if both air-inlets are blocked, it will take longer for the engine to reach starting temperatures.

7) When the engine is warm enough that the prop moves freely and the carburetor or fuel-injector is able to vaporize fuel correctly (about 45 minutes at -4° F/-20° C; 1+20 at -40° F/-40 C), the blowtorch(es) can be turned off and set aside to cool. At this point, the hot oil is poured into the engine, three or four shots of priming are given, and the engine is turned over four to six times by hand. It is now ready to go, and should start on the first turn of the prop.

* Propane torch method. Some pilots use a propane torch of the "weed-burner" type at base camp—they are not suitable for the survival kit. These torches (sold by Aircraft Components, Inc., as well as in hardware stores) have a long-handle, about ten feet of hose, and a needle-valve. They attach to regular refillable propane tanks (five-gallon or larger), which are available from local fuel suppliers. Costing between $50.00 and $60.00 (exclusive of the tank), they do the job, though you must be careful when using them not to apply too much heat to the engine compartment. Naturally, the fuel tank must be kept inside a heated building when not in use; otherwise, the cold will freeze the propane, preventing it from flowing properly.

* Coleman camp stoves. Along with old-fashioned plumber's pots, these are sometimes used, but are not all that good.

* The last combustion preheater I will mention is the Kennon SureStart II. This little unit comes in a toolbox and weighs 22 pounds. It has an electric blower that attaches to your airplane's battery, uses 14.1 oz. disposable propane cylinders, and costs about $500.00. As part of our throwaway culture, it is unquestionably cute, and I know of a flying school that uses them to warm their Cessna 152s on cold mornings. While not the sort of thing a bush pilot would want—consider being tied to city stores and their supply of little disposable cylinders—it is quite popular with private pilots who do not fly regularly in the winter, but need to occasionally heat their airplane either in a no-electricity hangar, or out at the tiedown. In these circumstances it does a good job.

Cockpit

While the engine compartment is the primary area we have to preheat—

for VFR flight it is the only area that is truly necessary—we do have to consider the cockpit. Besides comfort, the reasons for this are:

1) In airplanes that are to be operated at night or in IMC, including whiteouts, radios and gyros must be at or above their minimum operating temperatures. With panel-mounted radios and indicators — HSI, RMI, VOR, etc. — this is usually -4° F/-20° C; with pneumatic gyros, -22° F/ -30° C. (The slaved remote DG and flux detector that drive the King KI 525 HSI and KI 229 RMI are certified to -66° F/-54° C).

2) Windshield condensation can often be quite severe at very low temperatures, when defroster efficiency is impaired for the first few minutes of the flight. A little heat, just enough to remove cold-soaking, will go a long way towards resolving this problem.

I use one or two 1500 watt electric "car warmer" heaters in the cockpit. In the city, where electricity is available, I plug them in the night before, at the same time I plug in the engine preheater. In the bush, at my base of operations, I use a small electric generator to operate the heaters, turning them on at the same time I start preheating the engine with blowtorches.

In this way, by the time the plane is ready to go, the cockpit and instruments are warm enough to be operational.

(Tanis manufactures a panel instrument heater system, which I have never tried because I prefer heating the whole cockpit with a car warmer, but I imagine it works as well as the rest of their equipment.)

Away from my base, in the bush where no power is available, I have to content myself with VFR until the cabin heater has warmed the radios, indicators, and gyros to their operating temperatures. This usually works out okay, for low temperatures in the bush, away from the cities and their dense blankets of ice fog, usually mean VFR weather.

Battery

Cold also affects the battery; at seriously low temperatures it will not have enough power to start the engine. In the city, when electricity is available, many pilots with C-180/185s and Super Cubs will use a simple battery heater, such as the one sold by Tanis, or put a small "car-warmer" heater in the battery compartment. Pilots with airplanes like the C-206 do not need to bother, for the battery is in the engine compartment and will be kept warm by the engine heater as long as the engine cover is in place. In the bush, however, things are different, and you must either remove the battery and take it inside for the night so it can be kept warm, or you must handprop the airplane for the first flight of the following day.

Chapter 13

Cold Weather Engine Operations

Complacency is always the enemy, and it is all too easy during the warm, seemingly endless days of summer to forget that winter waits just around the corner. Then, one morning you wake to find a gentle frost decorating the trees and the thermometer reading 20° F/-7° C.

At this point you wake to reality and realize that winter, while not yet here, is certainly close at hand. It will not be long before the temperature spectrum is such that you must pay extra attention on a continuous basis to the operation of your engines. By cold weather, I mean temperatures between 30° F/-1° C and -35° F/-37° C. With temperatures above this operating spectrum, no precautions beyond the ordinary need be taken. With temperatures below this range, operations should be conducted only in emergencies, due to the stresses placed on engines and airframes.

The first thing you will want to do on arrival at your airplane is preheat its engine. While at the upper end of the cold weather operating spectrum you can usually manage to get fired up okay with a bit of over-priming and a little time spent leaning on the starter, this can cause five major problems:

1) The excess gasoline in the cylinders caused by overpriming will wash the thin coating of oil off their walls, increasing friction and therefore accelerating wear. This will lead to decreased TBO and added maintenance expenses at a time when you need all the help you can get to cut your operating expenses while still maintaining a high level of safety and dependability.

2) The different rates of expansion of the different metals when going from very cold to very hot—basically the aluminum pistons and the steel cylinders—will also increase wear, which will further reduce TBO, and therefore add to maintenance costs.

3) Grinding away on the starter, besides overheating it and thus leading to additional maintenance costs, will run down the battery, which in cold weather is more inefficient than it would be under normal conditions. If the battery discharges too much, you will need to get a booster battery, or hand-start (prop) the engine. If the airplane is sitting on ice, this can be interesting.

4) If the engine fires, coughs, sputters, and dies, the spark plugs can become frosted. This happens when the temperature in the cylinders rises rapidly above freezing, causing condensation to form, then drops back down again. Once this happens, you have the choice of removing the spark plugs, cleaning them, and trying again—usually a lost cause—or going through a total preheating of the engine.

5) With carburetor equipped engines, there is always the risk of a backfire causing an engine fire. If this happens while the engine is running, the first course of action is usually to try and suck the flames back into the engine by opening the throttle. If this does not work, you will then have to use the fire extinguisher. (A fire extinguisher *must* always be close to hand during a winter start, whether or not the engine has been preheated.) However, this type of situation is best avoided, for if the extinguisher needs to be used, it then becomes necessary to open up the cowling and carefully inspect the engine and all lines and wires for damage.

A little heat, often just enough to enable easy fuel vaporization, is, at the *upper* range of the temperature spectrum, all that is needed. At lower temperatures, of course, the job must be done right and the engine thoroughly preheated prior to attempting a start.

Winterization Kits

With the exception of the turbocharged 206, you must modify your airplanes for cold weather operations. The factory provides late model airplanes with convenient kits which are used below specifically stated temperatures. On the Super Cub, it is standard practice to drill a small hole in the breather tube near the exhaust system to protect the engine against moisture freezing and plugging the bottom of the tube—a situation which could easily lead to a blown prop seal and catastrophic loss of engine oil while in flight.

One bit of advice often encountered around the airport when cold weather rolls around is to use duct tape to partially block the air cooling inlets behind the prop. The idea is good; after all, the winter kits many manufacturers provide for their planes partially block the air flow over the cylinders. However, the factory's air-blocking plates are designed with a proper understanding of the airflow patterns over the cylinders so hot spots do not develop and reduce engine life. If you are going to do this type of thing, you would be better off taping the bottom rear of the cowling to slow the passage of air through the engine compartment. (If the airplane is a Super Cub, Atlee Dodge manufactures a cowl-flap kit for this purpose.) However, I must admit I have never felt it necessary to do this with the 150 horsepower Cubs, finding the hole in the breather tube and the factory oil cooler plate adequate. As to the Cessnas, the factory provides a winter kit for them, and it too is adequate.

Putting the Pilot in the Loop

Summer or winter, good operating practices should always be followed. In winter, however, this is especially important, for seriously cold temperatures make all operations more critical. These good operating practices are:

1) Throttle action. A major sign of the pro is the way he operates the throttle. A smooth, carefully paced application of power on takeoff; gentle, precise changes while setting up climb; careful, almost finicky, adjustment for cruise; and a slow, thoughtful, paced retardation for descent and landing—these are what you should strive for.

2) Mixture setting. Maximum engine longevity is highly influenced by fuel/air mix, so careful and precise use of the mixture control is very

important. Even an engine as bulletproof as the Super Cub's Lycoming 0-320-A2B benefits from careful and knowing mixture control, and, for the Continental engines in our Cessnas, it becomes critical. Most pilots use the mixture control as a way to lean the engine to increase range and help keep the plugs clean; I use it—along with MP, RPM, aircraft pitch attitude, and cowl flaps—as a means to adjust engine temperatures to their optimum.

In recent years, a sophisticated, computerized EGT/CHT/TIT gauge (manufactured by Insight) called the GEM has become available. This unit, with a column of orange light bars for each cylinder, and a numerical readout for TIT, is literally worth its weight in gold and pays for itself over and over again. Highly recommended.

3) Cylinder-head temperatures. A good way to shorten engine life is to push the nose over when ready to descend, and come back all the way on the power. Shock-cooling is a subject that gets debated a lot, but the bottom line is that your engine appreciates gradual temperature changes and will reward you with a much longer TBO if you do your part.

4) The approach. Power should be carried on approach, not only to assist in precision off-airport landings, but to keep from shock-cooling the engine. This is particularly important in cold weather, but it should be a part of your daily operations no matter what the OATs.

5) Shutting down. After landing, the engine should be allowed to idle for a couple of minutes so temperatures can stabilize prior to shutting down.

The above operating practices are important for engine longevity under all temperature conditions, but they become more and more critical the colder it gets. As you begin to approach the cold weather spectrum, however, you will have to expand your engine-management techniques even further.

Taxiing out, if you have a constant-speed prop, you will want to cycle it a few times to make sure the cold oil is moved out of the hub and the prop changes pitch freely. If you do not do this adequately, the prop will be prone to cycle during takeoff—a bad thing if you are operating from a short field, and inelegant at the best of times.

Once in the air, you will want to climb through the temperature inversion into warmer air. Sometimes it takes a bit of experimenting to locate the warmest level. Conditions permitting, you should remain in it over the route of your flight. This is one reason so many bush planes have thermometers taped to their wing struts where the pilot can easily see them. By letting the engine run at a more normal temperature, the warmer air will also improve the functioning of the cabin heater.

Naturally, to ensure more normal engine temperatures, you will use slightly higher-than-normal power settings while cruising. In warm weather I usually cruise at about 60% power, but in cold weather I usually cruise at 75%. By doing this, I am seeking high manifold pressures, so with a constant-speed prop I will choose the highest manifold pressure my POH will allow for my altitude and temperature, and I will then set the corresponding RPM. This is one of those small matters that seem to help a lot—summer and winter, my engines all easily manage to make TBO, even though they are used under the rough conditions of off-airport bush operations on wheels, skis, and floats.

Letdown from altitude is one of the most delicate parts of any flight, especially in winter, and great care must be taken under adverse temperature conditions to prevent damage, which may range from a cracked exhaust stack to the whole airplane, caused by engine failure. The most critical problems are:

Thermal shock. As noted earlier, coming back on the power abruptly and gliding down from altitude will cause the engine to cool too rapidly. The way to prevent this is to carry a large amount of power during letdown, which is somewhat easier said than done on many airplanes.

With the Super Cub on board skis, I begin my letdown far out, reducing the power only slightly (wheel-skis, because of their added drag, make this letdown much easier). While this also increases the time spent in the cold air below the inversion, there is nothing to do but bear it.

With a constant-speed prop and large flaps (like those on the C-180/185 and C-206), the situation is somewhat better:

1) Make your initial power reduction with RPM rather than MP. You are seeking the lowest RPM within the green arc that will permit smooth operation (lack of vibration).

2) Once below approach flap-limit speed, feed in 10° of flaps.

3) Now begin a timed MP reduction, but no faster than two inches every two minutes, and no more than necessary to obtain a 500 fpm rate of descent.

4) Lean your engine to either peak TIT (on turbocharged equipment) or peak EGT. Do not forget, however, to enrich the mixture enough to make an emergency go-around, in case one is required, prior to landing.

All this takes a bit of planning, but it is important. DME or Loran/GPS helps when out of familiar areas, by providing distance and time-to-station information.

Fuel vaporization. Cold air retards fuel vaporization, and very cold air can easily cause engine failure on approach. If your airplane has a carburetor-equipped engine, like those the Super Cub and Cessna 180 use, you will naturally use carburetor heat to assist vaporization and mixture distribution. Here is where the GEM really shines—it makes even subtle mixture changes obvious.

The major rule is, *in temperatures below -20° F/-29° C the throttle must never be closed until the field is safely made.* Approaches must always be planned with this rule in mind.

If it should ever become necessary under these conditions to add power once the throttle has been closed and the engine fails to respond, the best thing to do is retract the throttle all the way, then *very slowly* feed in a bit of fuel. I learned this lesson quite early in my bush flying career in a 125 horsepower ski-equipped Super Cub. I was making an emergency flight for a customer into Chitina one morning, with a better-to-stay-at-home surface temperature of -51° F/-46° C. My destination was the small, tree-lined lake on the edge of town known, with typical Alaskan flair, as Town Lake. The engine failure just above the ice—as a startled cow moose ran out from its hidden bed into our path—was impressive; the instinctive resolution of the problem enlightening; the subjective time-span until the engine caught awesome. My passenger never even knew the problem existed. I never forgot.

Section I, Part V

Navigation Tips: Silk Scarf to High Tech

Navigation Tips: Silk Scarf to High Tech

The North is a vast area, sparsely populated by people and navaids. As a result, there is not much of an airway system. While in recent years it has improved, it is still not capable of much more than taking us from one major community to another. This means that, in an area of widely scattered villages, cabins, and hunting/fishing/exploration camps, IFR flight is quite limited, and that VFR navigation is dependent not only on how *well* you know a given area, but also on the old-fashioned techniques of pilotage and dead reckoning.

Chapter 14 (Silk Scarf) takes a look at the techniques used by bush pilots since the 1930s. They have a long and honorable history behind them and are still in wide use today.

Chapter 15 (High Tech) takes a look at some of the new techniques based on modern Loran and Global Positioning System (GPS) technology. These new units, especially when interfaced with a Radio Magnetic Indicator (RMI), are the wave of the future.

While many, if not most, bush pilots limit themselves to VFR operations—shunning IFR as if it were the plague—I use both techniques. It is my opinion that, where appropriate, use of instruments provides a pronounced safety edge.

However, in this section I will focus mostly on VFR tips, the type of tips that are necessary for everyday bush flying. As for IFR tips, you will find them mentioned throughout this book wherever appropriate.

Chapter 14

Thumb on Map (Silk Scarf)

Well, you've got to start somewhere, and with navigation the somewhere is thumb on map. In order to appreciate how I learned my lessons, a little personal history is necessary.

As a kid, three things fascinated me: music (mostly Baroque and chamber), books, and airplanes. Unfortunately, they did not fascinate my parents, and so the classic struggle was born. The upshot was that I got an after-school job in a gas station, saved my money for a couple of years (although I was only fourteen I already had the capability to form long-range plans and put them into action), and near the end of my sixteenth year I ran away from home to San Diego where I rented a room and started flying lessons.

I was quickly tracked down, like any incompetent criminal, but not before I was just about ready to solo. My father, filled with trepidation, came down from Los Angeles to witness the great event. After I managed to drag the 65 hp "Airknocker" around the patch a few times and got my shirttail cut off, we struck a bargain. I would be permitted to remain in San Diego and get my private pilot's license if afterward I would return home and go back to school. My father was a good guy and doing his frustrated best to control his unruly son (small airplanes terrified him, as they do, to this day, everybody in my family), so I agreed and went on to get my license.

After I got my license, I returned to Los Angeles, but I did not forget about flying as my family for some reason had expected. When they found out that I had traded my car for an old PT-19 that I was fixing up and planning to fly, the you-know-what hit the fan.

I soon realized that flying wasn't going to work out, not with all the opposition I was getting; and I realized as well that the stress I was under was a definite hazard to safety. So, I struck another bargain: I promised to give up flying if the opposition to my studying music would cease. And so I took the low road, capitulated to superior force, and drowned my sorrows in Bach and Corelli.

Eventually I flew the coop, got married, and for a number of years toured Africa, the Middle East, and India as an itinerant cellist, making a very marginal but interesting living. Along the way, my wife gave birth to two children, one in Germany and the other in Ceylon.

Finally, after many strange adventures, and even stranger lands, we ended up in Alaska with a tent for a home, and I started looking around for some way to make a living. It was obvious that Alaska did not hold musicians in high esteem.

Well, the answer was not very difficult to find, not for a guy with my mindset. It was written, you might say, in the sky.

So, in my 29th year, I packed my family into an International Scout 4x4 I had managed to purchase and went looking for a Super Cub I could afford. In Red Bluff, California, I finally found what I wanted, an old ratty 125 hp Cub that I could have in trade for my car and $2,000 (life sometimes seems to work in cycles). The guy who owned it spent a couple of days checking me out in it—I had not touched the controls of a small airplane for 12 years—and it came to pass that on a cold January day I loaded my trusting family, kids and all, into the Super Cub (with survival gear piled all around them), and headed north.

Ah, youth… Looking back, I shudder contemplating what I did. On the trip back to Alaska, and for the first two years I owned that airplane, my rabbit's foot got so overworked that I wore it out. Today, there is hardly a hair left on it. But in the process I sure learned a lot.

Pilotage is primarily the art of navigating by landmarks. These landmarks can be roads, valleys, rivers, lakes, glaciers, mountains, anything that provides real-time data as to where you are. When using a chart, the usual method is to pick out the landmarks you will be using, along with compass headings and estimated time from one point to the next, and go for it. Pilotage got me safely to Alaska that first year, and to this day I use it when circumstances dictate. Let's take a look at the various landmarks and see how you will be using them in Northern Canada and Alaska.

* Roads. Flying in the Los Angeles basin, roads are not the best landmarks, but fly into the Yukon Territory or Alaska, and things become very different. Here you have a broad expanse of wilderness with few roads, and every one of them is an important navigational feature.

Take the Alaska Highway, for example. From Fort St. John it winds in lonely splendor along the edge of the mountains to Fort Nelson. There it enters them and winds through passes and narrow valleys to Watson Lake, Teslin, Whitehorse, and Burwash Landing, finally reaching Northway, Alaska (where you clear Customs).

On that first adventurous flight to Alaska in the 125 hp Super Cub, we followed the Alaska Highway from Fort St. John through an unforecast blizzard with visibilities at times less than a half-mile in snow squalls to Fort Nelson, where we spotted the road—clearly marked on the chart—to the airport. Turning right, we followed the road to its destination, landed, tied the airplane down, got a ride into town, and found a warm, comfortable motel (where we spent almost a week waiting out the storm).

All during that flight, there was never any question of where we were, or where we were headed, and if the weather got a bit too tough, well, the road was wide enough to land on almost anywhere, even for a greenhorn such as myself. It was much like flying over a runway hundreds of miles long.

All northern roads make good VFR routes. Used thoughtfully, ticking your landmarks off as you pass them, keeping track of time, you will never get lost, and will always have an emergency 'out' in case of need.

* Valleys. Valleys, too, make good routes. To go from Chitina—at the

confluence of the Copper and Chitina Rivers—to my place, you head up the Chitina Valley until you reach the Nizina Valley, then follow it until you come to the short, rough gravel bar I use as an airstrip. This gravel bar has, since 1968, been marked on the Anchorage Sectional Aeronautical Chart as FEPCO (Pvt).

* Mountain passes. For the VFR pilot, mountain passes are a way of life (see illustrations 32 and 33). Going from Anchorage to Iliamna VFR, you would follow the Lake Clark Pass; from Anchorage to Gulkana, Chickaloon Pass; from Bettles to Umiat, Anaktuvuk Pass; from Prince George, BC, to Watson Lake, YT, the long pass called The Trench.

All Northern passes have one characteristic in common: when the weather gets marginal outside the passes, it becomes dangerous in them. Fog, low clouds, and snow squalls are the most serious, though when the wind gets up, at times the ride has to be experienced to be believed.

I must admit I don't much like passes, and when I have the chance I fly above them, often IFR or above the clouds (see illustration 20). While I have flown Chickaloon Pass more times than I care to remember in VFR-only Super Cubs, I have never managed to make it through Lake Clark Pass VFR. Each time I have had to fly that route, the pass has been marginal to closed, and, being in an IFR airplane, I have climbed up to altitude, put on my oxygen mask, and shot the approach on the other side.

I know too many pilots who have died in passes to feel fond of flying them when the weather gets low.

* Rivers. Rivers are one of the great navigation aids in the north. Most villages and cabins are along them, and any major river will take you to civilization sooner or later. Large or small, rivers and creeks are your VFR highways through the bush (see Illustrations 34 and 35).

* Large lakes. Large lakes can be spotted from quite a distance when at altitude, and make good checkpoints when navigating by dead reckoning or pilotage (see Illustration 19).

* Coastline. The coast too is much like a river, with settlements, and, between Cape Lisburne and Cape Dyer, the long lonely chain of military DEWLINE stations.

When navigating through a mountain pass to a settlement on the coast, it is wise to plan your heading across the tundra so that you intercept the coast several miles to one side or the other of your destination. That way, unexpected wind drift will not get you so far off course that you get—as they call it in the north—"confused."

The rest of it is just getting to know your basic area of operations well, as well as you know your cabin or airplane. (See Chapter 24 for an example of this type of navigation during difficult weather.)

Chapter 15

The LORAN/GPS RMI Combo (High Tech)

When I started flying the bush, navaids, mostly NDBs, were few and far between, so it is not surprising that I developed a fondness for the radio magnetic indicator (RMI). But of what real use in an area rapidly converting to VORs is an expensive instrument that, for all practical purposes, takes a nice reasonable VOR and turns it into a NDB? One that, in fact, forces us to exchange our nice HSI or CDI—in a strange M. C. Escher transformation—for a bird-dog pointer just like an ADF's?

Well, a few years back, the answer would have been simple—not much. But that was before the current revolution in avionics, and the widespread development of sophisticated loran and emerging GPS technology. These new devices, with their interfaces and internal algorithms designed to take advantage of the RMI's unique capabilities, are resurrecting the RMI from the dark ages of instrument flying and bringing it to the forefront of modern navigation.

In the second part of this chapter I will describe some of the new uses for the RMI that the bush pilot will appreciate, but for now let's take a look at how the instrument operates and how the pilot flies it.

In most incarnations, the RMI has a slaved compass card, along with two needles, one for the ADF and the other for VORs or loran/GPS waypoints. The two needles always point TO the stations tuned, and your bearing TO or FROM the stations is found by reading the numbers under the head or tail of the needles. This makes your position relative to the stations and aircraft heading quite clear.

Most airplanes with RMIs and HSIs also have DME; thus, when the RMI is tuned to a VORTAC, you'll not only have bearing TO or FROM the station, but distance as well.

In its most basic application, the RMI is used to track TO or FROM a station, or to intercept a radial or bearing. To illustrate the principles involved, let's assume we're on a mid-winter vacation down south, and use a short flight out of Tucson, AZ along V-202 (see enroute chart #1). This airway departs TUS on the 107-degree radial, goes 25 miles to MESCA intersection, and intercepts the CIE 245-degree radial. The course, therefore, will be 107° to MESCA, then left to 065°.

For this demonstration flight, tune both navigation receivers to the TUS VOR, and place CIE's frequency in the standby windows. Also, switch the HSI and RMI to nav 1, and the number 2 VOR/LOC indicator to nav 2. Set the number 2 VOR/LOC indicator's OBS to 107°.

Departing Tucson, the 107-degree radial is intercepted off the end of the runway, and climbing through 5000 feet the compass cards on both the HSI, with the D-Bar centered, and RMI are showing 117°. This is our first interesting comparison of the HSI and RMI, for both instruments show the same offset for a southerly wind correction, and both needles show 107°. The head of the HSI needle is on 107 with the TO-FROM indicator showing FROM; and the tail of the RMI needle is also on 107, which indicates FROM.

As long as the D-Bar is kept centered, or the RMI needle is on 107, you

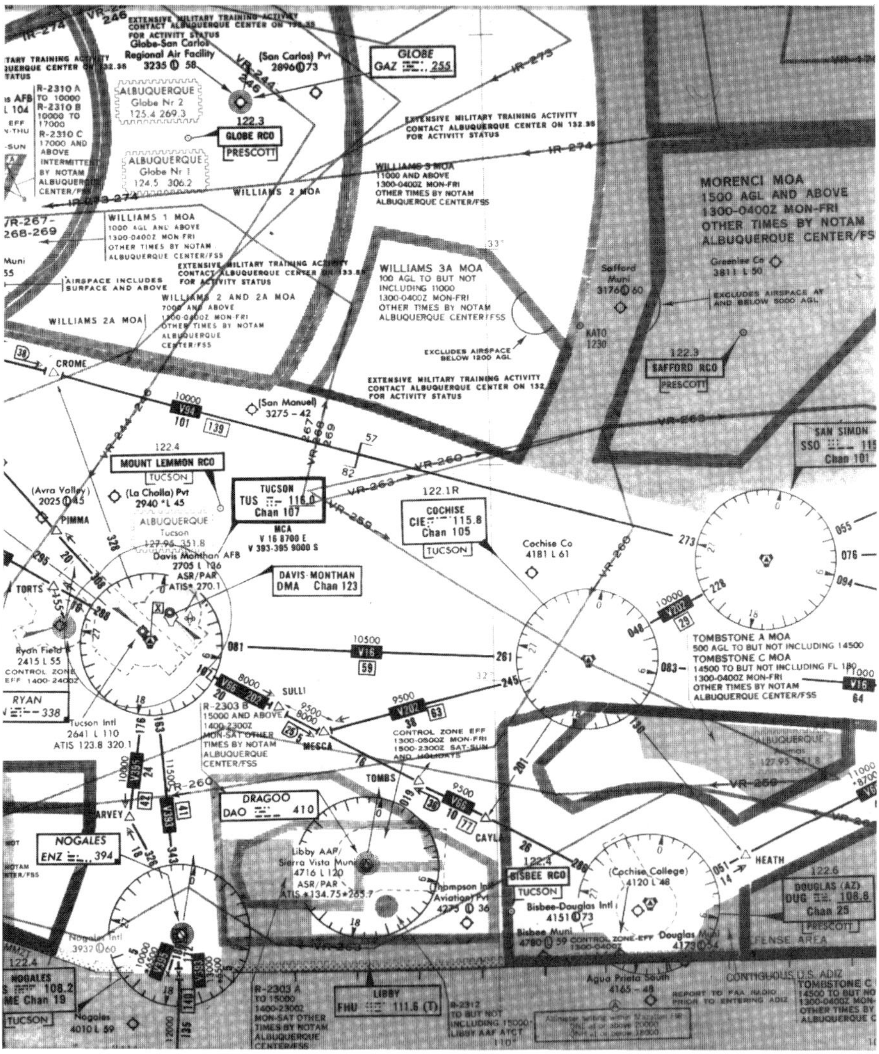

U. S. Government L-4 enroute low altitude chart showing the TUS-MESCA-CIE portion of V202.

Not to be used for navigational purposes.

will be tracking the correct radial. Heading will naturally shift as the fickle wind does, but everything else will remain the same.

Now put your autopilot (if you have one) in heading mode and use nav 2 to continue tracking the TUS 107-degree radial while you switch nav 1 to CIE and dial in the 245-degree radial on your HSI. Passing the 12 DME (TUS) point, the D-Bar on the HSI will begin to come alive and move off the stop. The RMI will show approximately 072° under the head of the needle, which is the present bearing TO the station.

The bearing FROM the station, which is shown under the tail of the needle, would be the reciprocal: 252°. At half-scale deflection on the HSI, the RMI will show approximately 068°. Both instruments will still indicate a magnetic heading of 117°.

As the target radial is intercepted, three things will happen: the TUS DME will show 25 miles, the HSI D-Bar will center, and the RMI needle will show 065° under its head (and the 245 degree reciprocal—which is the radial we're on—under its tail).

To fly to the station on the 245-degree radial, all you have to do is keep the D-Bar centered on the HSI, or 065 under the head of the RMI needle. In a no-wind condition, a heading of 065° will line up with both the HSI and the RMI needles.

However, no-wind conditions are rare, so compensation is required. With the HSI, if wind starts to drift you off course while tracking TO or FROM a station, you simply turn toward the D-Bar offset. Then it's just a matter of cut-and-try until you find the magnetic heading that will keep the D-Bar happily centered.

With the RMI, things are slightly different (though no more difficult), and this is the point where most pilots new to an RMI have their initial problems with it. When tracking TO a station, if wind drifts you off course, the head of the RMI needle will point the way to turn to get back on course. In other words, if your RMI needle moves off the bearing you want because of wind, all you need do is turn toward the direction the needle points in order to get back on course.

Once the proper number is back under the head of the needle (in this case, 065°), all you need do is find the magnetic heading that will keep the needle where you want it, just as you'd do with the HSI. Keep in mind that since you are tracking TO the station (and this is a very important point), the head of the RMI needle will be pointing toward the nose of the airplane.

However, when tracking FROM a station—and this too is very important—the head of the RMI needle will be pointing toward the tail of the airplane. Yet here, too, all you need do is turn in the direction the head of the needle indicates to get back on course: head of needle on the right, course to the right; head of needle on the left, course to the left.

Now that the basics of RMI use have been covered, let's go back up to Alaska and take a look at how the instrument is used for a DME arc approach with an NDB segment.

In the more interesting corners of the world, instrument approaches tend to be a bit interesting too, and the Sparrevohn NDB/VOR DME 34 (see approach plate #1) is no exception. Admittedly a bit of a job using just an HSI and ADF, it turns into a piece of cake with an RMI. All you need do is get yourself to the initial approach fix (IAF), make a 90-degree turn, then fly the arc.

The turn will put the RMI VOR needle directly off your right wingtip, where it will immediately start moving toward the tail. Once the needle has moved 10° toward the tail, turn 20° towards the station. If the wind isn't too horrible, this will put the needle 10° ahead of the wingtip, which will once again begin to move toward the tail. Continue this process, keeping a sharp eye on your DME to make sure you don't get too far off the arc. Two tenths of a mile on the outside is about as far as you should allow yourself to stray, though slightly more on the inside is permissible.

All this time, of course, you would be making the first stepdown, while monitoring your RMI ADF needle. When the ADF needle reaches 347°, turn right and track directly to the CRN beacon, at the same time making your next stepdown. The final stepdown is at the 5.5 DME fix. It's good practice to use available cross checks; as a cross check for this fix, you'd have the SQA 148-degree radial. At this point, descend to the MDA. It's important to keep the ADF needle right on target (the RMI VOR needle will be defining your missed approach point for you). If the ADF needle drifts off more than a couple of degrees, execute a missed approach, which has to be flown very promptly and carefully, as the approach plate warns.

So here we have the basics of RMI use in its traditional guise: using it enroute TO and FROM various stations, intercepting radials and bearings,

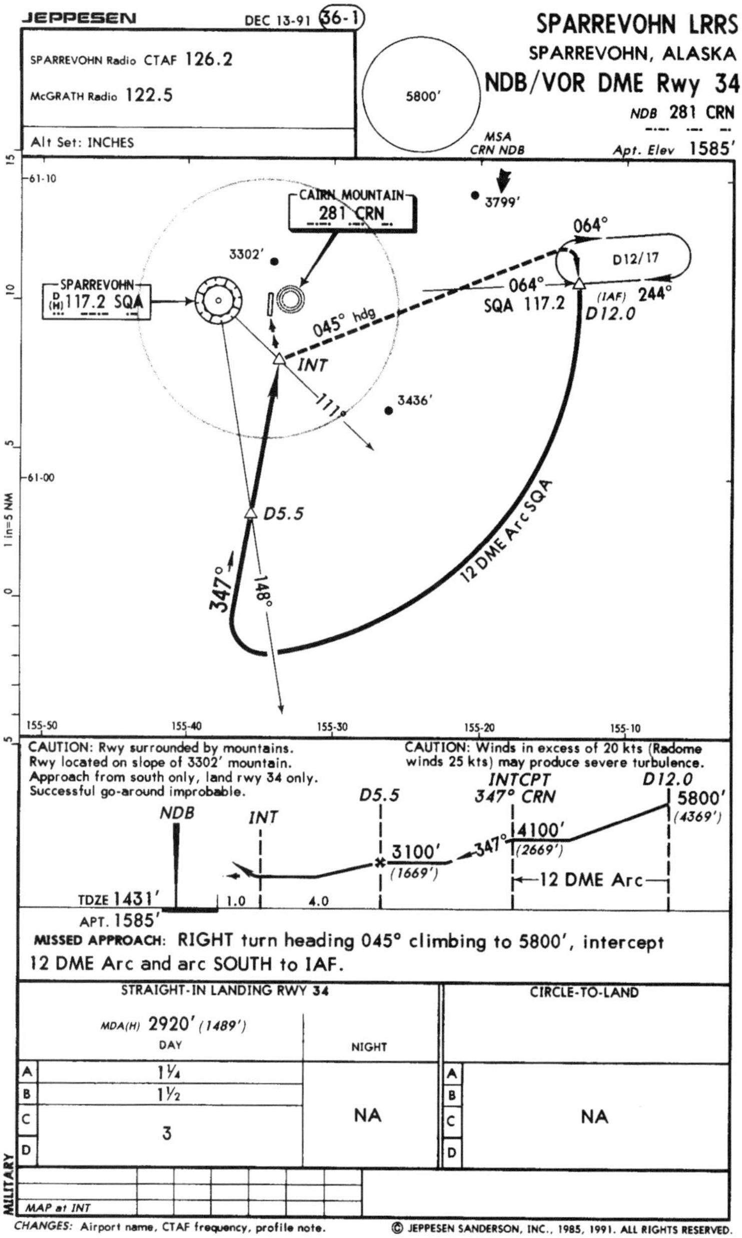

Sparrevohn NDB/VOR DME Rwy 34 approach chart.

Reproduced with permission of Jeppesen Sanderson, Inc. Not to be used for navigational purposes.

and flying approaches—NDB, VOR, VOR/DME arc and the occasional cuties like Sparrevohn's, which really demand its use. Yet the RMI's real utility, and, for me, its primary interest, lies in its new high tech applications.

Interface: the LORAN/GPS-driven RMI

We are moving into an age of VOR/NDB independent navigation, one in which the RMI is being put to increasing use in loran, and shortly, GPS, navigation. At the time this chapter is being written, the only IFR-certified loran that has provisions for an RMI interface is the King KLN-88, which I have in my Cessna Turbo 206. That's the model I'll use as an example here, but more loran and GPS units will be available soon, and the most useful models will have the RMI interface.

* The phantom LOM. In this day of radar vectors, we often find ourselves flying hither and yon for spacing purposes. When in IMC, it's easy to get disoriented and end up with a bad vector that puts us high over the marker with a "cleared for the approach."

As the FAA has been making like busy beavers decommissioning LOMs, we have to take matters into our own hands. This is where the KLN-88/RMI combo really shines. For continuous position awareness relative to the OM, enter the marker's identifier in your loran's active window and switch your RMI to it. Then, as you get the grand tour around the countryside, you will easily be able to keep precise track of your bearing and distance to the marker. No more unpleasant surprises. If you should get a bad vector, you will know what to do.

* Great-circle loran navigation. Loran direct is a great way to go, for you are able to avoid all the zigs and zags of the VOR airways. But loran always needs a cross check to make sure all is going well, especially when IFR, and here is where the RMI is useful.

With the HSI giving on-course guidance, and perhaps driving the autopilot as well, call up the KLN-88's nav 2 page. This page will provide you with the distance and radial from the nearest VOR. Then switch your RMI to either of your VOR receivers, and both it and your DME to the indicated VORTAC. This will provide a running cross check of your loran track, and, at a glance, you'll be able to determine the quality of your loran's data output. When IFR enroute, this is the RMI/loran feature I use most.

* Find that airport. On hazy "VFR" days—of which there are too many—it helps to know where an airport is, especially if you are a stranger to the area. So enter the airport's identifier into the loran's active window, and let the RMI point to it. No matter where you are, or how low, the RMI needle will always show you the bearing TO the airport, while the loran shows the distance.

* Find that hunting/fishing/mining/expedition camp. In Alaska and Northern Canada, you often have to fly VFR under a very low ceiling over obscure tundra or forest terrain to a destination that's many miles away. Sometimes terrain or weather will require numerous course deviations. Under these conditions, without loran/GPS RMI assistance, you could easily get "disoriented."

While this type of navigation to a waypoint can be done with a CDI, it's better managed with an RMI, for the RMI gives you your navigational picture at a glance, while the CDI (or HSI) requires you to keep changing instrument settings. Flying low-level because of difficult weather is no time to be messing with the radios.

With the new loran/RMI interfaces, it's obvious that the only limit on how a pilot utilizes this equipment will be his imagination. Strings of waypoints around MOAs and other restricted areas, complex STARs to an IAF in difficult terrain, keeping an eye on ATC: it's all easy to accomplish.

And once entered into the computer, following the charted waypoints is as simple and straightforward as can be. Using this system in my daily flying for the past few years, I have come to depend on and trust it, and the loran/RMI combo has become my favorite instrument combination.

It will remain that way until the IFR-certified GPS/RMI combo becomes a practical reality.

* * *

As this book was going to press, I installed a King KLN-90 GPS with RMI interface in the B36TC. My initial impressions are that GPS is far more accurate, and noticeably more solid, than Loran. I would suggest that those contemplating the purchase of a Loran get GPS instead. It is the wave of the future, for the VOR/DME system (and Loran with it) will soon join the old aural LF/MF ranges as a quaint and nostalgic part of aviation history.

Section II

Flying Techniques

Caution—this section contains *advanced techniques* that could be hazardous to nonproficient or poorly-trained pilots. Refer to *Disclaimer and Warning* on page v prior to attempting any maneuvers described in this book.

Section II, Part I

Wheels

Wheels

Bush pilots are, to put it mildly, an opinionated breed, so it is not unexpected to find that strong preferences exist in a profession that utilizes three different types of landing gear. Some pilots love floats, and feel the only true bush planes are those that operate from water. Others like skis, feeling that only with them is the epitome of northern bush flying truly manifested. I must confess to being a trifle stodgy—I prefer wheels.

Much of the basis behind these choices lies in the realms of personality and psychology. For example, I enjoy precision in flying over all else, and this is best realized on wheels when using very short strips. Some gravel bars, especially those used for hunting, are so critical in respect to length that the margin for error is in the neighborhood of ten feet.

Yet, how can you obtain the skills necessary to operate on these marginal strips? And, what can you do to hurry the learning process?

Well, starting with the time element, the unfortunate truth of the matter is that these skills can only be developed over a period that spans not days or weeks but many seasons. The mind is capable of grasping ideas fairly quickly, but it takes time for the skills to filter down to a deep enough level in the subconscious that proficiency finally becomes solid and ingrained.

While I was forced to master the basics of off-airport operations quickly, since a rough 500-foot gravel bar in front of my cabin was all I had for an airstrip (see illustration 8), this is not the optimum way to learn this skill. Far better to work up to serious short field operations slowly. So, I am going to provide a few exercises in this part of the book that might help you learn the trade a bit more safely, if not any more rapidly.

Becoming really proficient at anything requires an enormous amount of dedication, combined with work and study and time. The most productive attitude you can take towards this situation is to acknowledge that it is a career-long project that produces a new bit of information or insight on almost every flight. It is a process of ever-increasing refinement of detail, of learning how to recognize and remaining ahead of all the various trends that develop.

Perhaps this is one of the reasons I continue, even after all these years, to find flying so fascinating.

Chapter 16

Takeoffs

To use the most critical strips on wheels requires precision in three main areas:

1) During the takeoff run, aircraft attitude must be maintained within two degrees of optimum no matter how rough the airstrip.

2) During climbout, airspeed must be maintained within one knot of optimum for current conditions *until clear of obstacles.*

3) On approach to landing, the airspeed must be maintained within one knot of optimum.

Any deviation from the above figures will increase the runway and obstacle clearance required.

Specific figures as to airspeed and attitude required vary from flight to flight as operational and environmental conditions change. The trick, naturally, is not only in maintaining them, but in knowing what they are.

It is generally assumed that landing is the one area of the flight where the skill of the pilot shows most obviously, and when we wish to impress the non-pilot—or the low-time pilot, which is much the same thing—we make a point of "greasing" it on, knowing that he considers this the be-all and end-all of flight technique.

This little game is harmless, and serves a point in professional operations since it keeps the customer happy and reassured without any effort on your part; but you must take care not to become a believer in your own mystique. Studies of Northern accident patterns, along with experience, demonstrates that the reality of the matter is somewhat more complicated.

There is little question among pros that the takeoff phase of a flight can be tricky, and this shows up on the really short, rough, dogleg strips. Here, two basic building blocks determine your success or failure:

1) Load; and

2) Precision of technique.

On the matter of load, many airplanes require markedly more ground run for takeoff than landing, and this is especially true of the C-180/185. With the C-185 the book figures, for example, show that—at gross weight—the factory pilots can perform a landing with a ground run of only 480 feet, but they require 770 feet to get off again.

It is easy to see where this can lead you, and it should be noted that the factory figures are for a paved runway. With rough gravel strips this takeoff figure will be increased even more, and the spread between it and the landing figure will become additionally pronounced.

This is where extreme precision in technique comes in, for many gravel bar strips have variable surface textures along their length, and these textures require different angles of attack during different points in the takeoff run to obtain maximum performance (see illustration 16). And to this you must add all the other details that go into a properly executed departure.

Wind, density altitude, and load

It is important, initially, to take a look at this problem from two opposing points of view:

1) The pilot's view, which is aerodynamic considerations. Here we have the major concern of the mature pro: Bring 'em back alive—along with the airplane and all cargo. This takes precedence over all other considerations.

2) The boss's view, which is customer relations. The boss must placate the customer, who, because of economic considerations, naturally wishes to stuff as much into each flight as possible. Most customers have a constitutional inability to comprehend the equation of weight versus length and condition of strip. To them, if it'll fit inside the airplane, the airplane (just like their car) will carry it.

These two forces, if not brought into balance with the utmost care, will appear on our safety records as "pilot error"—which it is—though job security sometimes enters into the picture with a few of the less reputable Part 135 operators.

Knowledge is our answer here, so:

Out in the bush, with your airplane sitting on a small gravel bar next to a large pile of equipment and supplies, and your customers pressuring you to add just one more box to the load, a fine sense of where to draw the line as to what you can carry off the strip *under current conditions* is well worth having. One question is: How can you tell when you have reached the point of overload? A second is: How can you convey this information to your customers in a form they can understand and appreciate?

In town, with your airplane sitting on a paved airport with a FSS and Denalt *(den*sity *alt*itude) computer handy, the solution is easy. All you have to do is compute actual density altitude, find the proper multiplier, look in the POH, and have a usable answer that even contains a mild fudge-factor.

This not only provides you with a "book-figure" for the customers, but also provides a bit of guidance for your own use under difficult conditions. There is little doubt that it can sometimes be a handy gadget—even if only on the psychological level—and more than one bush pilot over the years has wished that such a gadget, properly programmed for bush working conditions, would show up on the market.

Let's indulge for a moment in a little dream about a computer, much like the Psion-based Flightmaster computer Mike Busch developed for navigational purposes, that the FAA, in its wisdom, is going to provide us in the name of safety. In it, we would be able to enter the following important factors:

* Weight and balance characteristics of the airplane, including empty weight, fuel on board, and pilot's weight;
* Length of strip;
* Gradient (slope) of strip;
* Condition of strip;
* Degree and position of dogleg (if any);
* Obstructions—trees, the base of a cliff, that moose that's waiting to run out in front of the airplane five knots before V_{lof};
* Temperature, humidity, pressure, and altitude, i.e., true density altitude;
* Wind direction and speed, including random gust factors from the surface to 500 feet above any obstacles;
* Condition of the airplane, engine, and prop;
* Psychological pressures and fatigue;
* Your flying ability *at that particular point in time;* and then, with the push of a button, come up with the answer: 327 lbs between stations 50-108.

Then, with your customers staring at the official readout on your FAA takeoff computer, and your mind at ease, you could weigh the cargo, load it, and head on your way, clearing the obstacles at the end of the strip by a comfortable two-foot margin as they look on in amazed and contented appreciation.

But...reality once again intrudes, and coming back to earth, you discover that your takeoff computer with its nice readouts for everybody to see does not exist yet, and you left your scale at home. What to do? How does the experienced bush pilot handle the situation? How does he manage to load the airplane and clear the trees by that two-foot margin while keeping his customers happy?

Well, the key is in the word *"experience."*

For what you do, of course, is learn to develop a "sixth sense" towards the problem of weight and balance and the shifting conditions of your airstrips just as you do towards the shifting patterns of the weather or the shifting approach paths required to make proper off-airport landings.

This sixth sense is quite easy to develop, and with a reasonable amount of application improves fairly rapidly. Part of it will come from your daily flying activities, and part of it will come from a formal program of study.

And, along with this will come the self-confidence and knowledge that will allow you to painlessly handle your customers when they say, "Uh, you got room for one more keg of nails?"

The way to develop this sixth sense is:

On a daily basis, with every takeoff from every airstrip, make an estimate prior to applying power of where you "feel" the airplane will lift off. This estimate can be made using any technique or combination of techniques that seem appropriate, as long as they attempt to take into consideration all the factors you would enter into the imaginary FAA takeoff computer.

During each takeoff, carefully monitor the feel of the airplane (seat-of-the-pants; shifting control pressures), and pay close attention to where and how these sensations relate spatially to your estimate of the liftoff point. *This is very important,* for these shifting relationships are not only the key to precisely determining takeoff requirements under varying conditions, but will also allow you to pinpoint possible problems or miscalculations prior to reaching the point of no return (or, in the terminology of the multi-engine boys, the accelerate-stop distance).

Your progress with this will at first be slightly erratic, and for a while you will find that the kindest thing that can be said is that your estimates are somewhat imprecise. However, as time goes on, you will begin to notice a very definite improvement, and eventually you will discover that you are hitting it within a fifty foot (Super Cub) to seventy-five foot (C-180/185, Beaver) "window."

Hopefully, this process will by now have become an automatic part of every departure, since continuing it for the rest of your bush-flying career is the only way you can hone and maintain your skills to that level where you would be willing to bet your life on them—which is what you will be doing if you wish to use the most critical landing areas.

From here you can proceed to your formal training. This training is designed to develop your skills to a professional level, and with proper application it will do so—providing you have the basic natural talents to build on.

As a criterion, using a Super Cub, the best pros can judge their load and distance to within ten pounds and ten feet; some also manage to do this with the C-180/185, and these are the pilots you find taking their lightly-loaded C-180/185s into strips that are so short they restrict most others to Super Cubs.

For your formal training, you will need a place to practice and experiment (see illustration 16 for an example of the type of strip that provides the best training). This place must have the following attributes:

* It must be dirt or gravel, since off-airport operations are performed on this type of terrain and the feel is quite different than pavement.

* It must be narrow for the following two reasons:

1) There is a marked visual shift in the relationships used for perception on this type of strip compared with regular airport runways.
2) You must train for the precision required in tracking within narrow limits.

* The strip must have good approaches, for your purpose here is not obstacle training but rather ground-run awareness.

* It should be long enough to allow gross weight operations with a comfortable safety margin.

* It should be close enough to your home base so you can get to it in all different types of flying weather for your experiments.

* It should be in a quiet area without other traffic or observers present so the job at hand will get your full attention.

Arriving at the practice strip with several five-gallon cans of fuel and a filter funnel, park the airplane at the end and get out to look things over from the ground. This ground inspection is one of the key elements of off-airport operations and should be performed on every bush strip you use, *every time you use it.* It is only by doing this that you will become aware of the many subtle factors at work through the changing seasons.

Strolling up and down the strip, feel the wind and temperature and note the rough and soft spots—these will change with use, and it will be necessary to file these changes in your mind as they occur. Pace off the distance to various landmarks along its length—rocks, bushes, etc. Then plan emergency ground tracks off the strip which, if available, will assist in minimizing damage in case of need.

Once this has been accomplished, you should return to your airplane and, with it empty except for a half-hour's worth of fuel, perform a normal short-field takeoff. This should be done with a smooth application of power, a slightly tail-low attitude—whatever is required for the current terrain conditions—and by allowing the airplane to lift off when ready.

You should carefully note where you first feel that unmistakable sensation in the seat of your pants that tells that the airplane is getting ready to fly, and *how much farther it is before the airplane actually lifts off.* This is very important and is part of your accelerate-stop computation.

Repeat this process over and over with the same load. Pay particular attention to the little differences that slight variations of technique produce. What you are looking for here is the technique that, for these conditions, produces the shortest ground run into ground effect.

Once you have this load configuration figured out, add a measured amount of weight at the forward limit of your c.g. and give that a try. Then shift the weight to your midrange c.g., and try that. Then to full aft c.g., always watching your balance envelope.

From here go to the next increase in weight, and then the next, until you reach your maximum allowable gross weight, and for the smart bush operator this is the final limit no matter what anybody says.

Then repeat this process from the beginning adding steadily increasing quantities of fuel, until you have a good feel for what payloads you can carry with different fuel loads.

Once you have a feel for these different weight and balance conditions on the practice strip under a variety of weather conditions, it will be time to begin your study of the more advanced and precise techniques described in the rest of this chapter.

Application of power

Often the first hint of a pilot's skill comes with the starting of the engine and the manner in which he applies power.

Smooth and careful precision is what we are looking for, and the surest sign of the duffer is in throttle technique, whether in the air or on the ground. According to Lycoming, misuse of the throttle in all its various forms is one of the two major reasons behind premature overhaul of airplane engines, the other being the internal buildup of moisture and corrosive acids due to under-utilization—a problem the bush pilot rarely has to face.

Misuse of the throttle also causes damage to the propeller and, irrespective of common belief, winding a non-turbocharged engine up to full power with the brakes locked is unnecessary on most bush strips and should be reserved for those that are so short that the advantage you need can be measured in five or ten feet. For the truth of the matter is that the propeller, being an airfoil corkscrewing through the atmosphere, requires a certain amount of forward motion before it will start to operate efficiently.

You should, for economic reasons, put this to the test, and then you will be able to determine exactly how much advantage it provides by measuring it against the normal-application-of-power lift-off point on your practice airstrip.

From a psychological point of view, it is easier on the customers if the job has a casual air about it, and this is another factor in favor of the smooth application of power on bush strips whenever possible.

Raising the tail

At rest, the tailwheel aircraft, especially on high-flotation tires or high-pedestal skis (see illustrations 6 and 17), has a high angle of attack, and this creates a considerable amount of aerodynamic drag as speed starts to build during the takeoff run. To neutralize this drag, raise the tail to the ground-run attitude as soon as possible. This requires a nice touch, for if you attempt to raise it too soon you will add to the drag; if you wait too long the additional drag will extend your run.

The best technique, especially with the C-180/185, is to allow the tail to remain on the ground until there is enough airflow over the horizontal stabilizer to balance its weight without introducing additional drag through a pronounced degree of down-elevator. At this point you will be able to raise the tail with a subtle amount of forward pressure, and from here on out select the attitude that will provide the maximum acceleration through liftoff.

Attitude (ground-run)

Your takeoff attitude will be basically determined by the surface conditions of your airstrip. With rough or soft strips, keep the tail somewhat low to provide a light tire footprint. With hard, smooth strips, bring it up a bit higher to take advantage of reduced frontal-area drag. If the ground is soft in some spots and hard in others, you might require a combination of different attitudes as your run progresses to obtain the best acceleration profile. And, of course, if there is much of a wind you will need an attitude that will keep the wheels firmly on the ground until you are ready for liftoff.

Use of flaps

It is common practice, on normal strips with smooth terrain, to set in about half-flaps prior to starting the ground run. This provides a smooth, uncomplicated liftoff that is comfortable for the customers and can be used with fairly short strips.

But on the tough strips—the short, soft, rutted ones—you must use a different technique. This technique is the main reason bush pilots prefer mechanical flaps operated with a Johnson bar (Super Cub; C-180/185) to either electric (C-206, etc.) or hydraulic (Beaver) flaps.

To put it briefly, the airplane is lifted into ground effect with flaps; then, while still in ground effect, it is allowed to accelerate to its best angle of climb speed, whereupon climb-out is initiated.

The trade-off here is decreased ground run for increased pilot skill, and, needless to say, this is a pro's technique that has the potential to get the amateur who does not understand how it works into trouble.

The initial problem will often show up in ground tracking, and is caused in part by distraction of attention, different control feel, and aerodynamic factors. This problem is quickly resolved, and many pilots do not even encounter it.

The main problem will come from lift-to-drag ratios. The degree of understanding you have determines the success of the maneuver; complete lack of understanding leads to operation behind the power curve with the resultant risk of "mushing" into the trees.

As you know, every airplane has a lift/drag ratio that varies with angle of attack (in and out of ground effect), flap configuration, airplane weight and balance, and a number of other factors such as correctness of rigging. This ratio is constantly changing in relation to the above factors in subtle and not-so-subtle ways from flight to flight.

Fortunately, this problem is nowhere near as formidable as it might appear, for you can feel the lift curve meet the drag curve in a seat-of-the-pants way as you apply your flaps, and because of this it is a simple matter to determine when you have reached the point of maximum lift for current conditions.

Because the C-180/185, due to its large flaps and greater weight, is not only more critical of technique than the Super Cub, but also far more rewarding in the degree of additional performance to be gained, it will be used for the operational example.

This is performed "by the numbers."

1) Set in 10° of flaps. This is not done for any aerodynamic reasons, but so you can reach the flap handle buried down on the floor without losing sight of the runway during the takeoff run. If your arms are long enough, you can leave the flaps up.
2) Apply power.
3) When the airflow over the horizontal stabilizer is capable of holding the weight of the tail, allow it to come up to ground-run attitude.
4) Shortly thereafter you will detect that distinct feeling that lets you know the airplane is getting ready to fly. Just beyond this point reach down and apply flaps.
5) As you pull the flap handle, you will feel the airplane lift free of the surface into ground effect. Its attitude will remain within a few degrees of what it was during the period of ground run just prior to flap application.
6) As you continue to apply greater degrees of flaps, over a one to two second time frame, you will shortly feel the drag curve meet the lift curve. At this point slightly relax your flap setting to the point of maximum lift.
7) Keeping the airplane flying level in ground-effect, slowly bleed off the flaps as it accelerates to best angle of climb speed.
8) If done properly, you will reach your best angle of climb flap setting at the same time you reach your best angle of climb speed. At this point, increase your angle of attack to nail this speed and leave ground-effect. You should maintain your best angle of climb until clear of any obstacles, and then reduce pitch to best rate of climb speed (or less, if desired) until reaching cruising altitude.

A simplified form of this technique can be used during the study period with the C-180/185. Here, just introduce 20° of flaps at the lift-off point and retain this setting until leaving ground-effect. While this isn't quite as efficient as the previously described technique, it does provide better performance than the standard set-in-the-flaps-and-go method while giving one a gradual introduction to the various forces being used.

With the Super Cub you are unfortunately limited in technique to leaving the flaps at 0° until the lift-off point, then introducing full flaps—which must be maintained until past any obstacles requiring best angle of climb. This is due to the fairly inefficient flaps, which produce the shortest takeoff and best angle of climb with one setting, irrespective of load or other conditions.

One final note on this technique: On really short fields you should not try to lift the airplane off as soon as possible but rather hold it on until reaching the very end of the strip. This is to guard against lifting off too soon, with the resultant risk of settling into the ground again due to a miscalculation or random gust.

Dogleg strips

Many bush strips are curved, because they are gravel bars created by water erosion. As a result, you will have to be able to land and takeoff in directions other than a straight line if you wish to use them.

The main thing you must learn is how much speed you can carry around a given curve before centrifugal force moves you off your desired track. This knowledge comes with experience.

There are two techniques in general use for takeoff:

1) Pacing. With this technique—preferred on the more comfortable strips—you apply power carefully to control acceleration so you arrive at the curve at just the right speed to go around it without drifting off to the side.

2) Power-and-brakes. This technique is used on the critical strips, and with it you go around the curve with full—or close to full—power and control groundspeed with brakes.

Both these techniques, like most of the ones in this book, require a nice feel for the airplane along with thoughtful practice to be mastered. And, of course, when performing them, you must continue to calculate your liftoff spot in advance.

90 and 180 degree turns directly after liftoff

This type of turn is often necessary to avoid obstructions when using bush strips, and in tight spots must be performed just out of ground effect.

It is done as follows:

1) Start the takeoff run in the normal manner.
2) When best angle of climb speed has been obtained climb out of ground effect to an altitude high enough to prevent dragging a wing tip.
3) Level off with a flap setting of 20° (C-180/185) and allow the airplane to accelerate to a speed five knots above the 60 degree angle-of-bank stalling speed as shown in the POH.
4) When this speed has been obtained, roll into the turn using no more than a 50 degree angle-of-bank. Be sure to maintain a level altitude until reaching the desired heading.
5) Once this heading has been achieved, accelerate to best rate of climb speed as you raise your flaps, adjust angle of attack to nail it, and climb to your cruising altitude.

(While this turn has been described in five separate steps, in practice it is done as one smooth, flowing maneuver, each step blending into the next in a seamless whole.)

Chapter 17

Approaches

In the cities, we have standard landing patterns, usually beginning with an 800 foot AGL left downwind, then a descent on base and final to, or just slightly beyond, the numbers. This pattern works well for the mix of airplanes usually found around airports, and provides a degree of predictability that is not only useful for consistently safe (and mostly satisfactory) landings but also provides order to the often saturated traffic situation.

In the bush, approaches to landing areas—because of the all-too-often difficult terrain considerations—are rarely able to follow the standard FAA traffic pattern. They defy rigid rules, and require of the pilot an ability to evaluate the terrain, the weather, and the performance capabilities of the airplane/pilot combination under real-time conditions.

Because of the many factors that go into a proper bush approach, let's take a bird's-eye view of the process.

Images

The Raven

He flew past it at 50 knots and half-flaps, checking, and it was as he remembered: 340 feet of narrow, doglegged gravel bar, surrounded by trees and brush in a deep canyon. Along one side the rocky creek from the glacier flowed; once, long ago, during a period of high water, it had created this landing area; eventually, it would destroy it. But for now the gravel bar was still usable, which was fortunate, for it was the only place within miles of where his customer needed to go that an airplane could be set down.

In Alaska, you use what you have and are thankful it is available. For here, a true airport is considered almost a decadent luxury.

"Do you think it's okay?" his passenger called worriedly from the rear seat of the Super Cub. "Looks pretty short to me."

The pilot, looking at the strip, nodded; and, as he did, a picture formed within his mind—the approach path, with the various "windows" along it, necessary to bring him to the threshold at the speed and altitude necessary for a safe landing. "Yeah," he agreed. "It's short—and rough. But good enough. We're going in."

He made his turn, headed downwind, maintaining his 50 knots at an altitude of about 20 feet above the trees; then, on coming abeam the point where he needed to enter the creekbed and descend below the trees, he turned once again, heading back on an angle. As he reached the edge of the forest, he smoothly came back on the power, carefully maintaining airspeed with attitude (elevator), and began his descent.

The creekbed twisted like a snake, its width roughly twice the airplane's wingspan, and the Super Cub followed, going lower and lower, airspeed locked on target, subtle alterations of power controlling descent. Along the edges, in patches of sand, were the tracks of moose and bear, clearly visible to the passenger, and above them a raven circled. Then they were going *around the final bend, this one narrower, and as they did, while in the turn, the pilot applied full flaps and slowed to 1.1 V_{SO}, using his sensitivity of feel for flying on the very edge of the stall, a high level of skill in this area of the envelope being one of the primary tools of the bush pilot's trade.*

In front of them the strip now lay, visible for the first time since they began the approach, the curve of its dogleg obvious, thick brush at its far end. As they neared the threshold and entered ground effect, the pilot raised the nose slightly, while smoothly adding a touch of power, and the Super Cub slowed even more.

Then the main gear was crossing the edge of the shallow bank, hardly a foot above its surface, and, as it did, the airplane's tail still over water, the pilot came back on the power and the Super Cub settled, touching gently, its large, soft, high-flotation tires now rolling smoothly over the rocks. Then the flaps were off—the brakes, as is proper bush technique, having been partially engaged just prior to touchdown—and they were tracking around the dogleg, coming to an easy stop. In front of them, close, but not too close for comfort, the brush waited, a common hazard in the bush. For, as is often the case with off-airport landing areas in Alaska, this was a one-way strip.

And had the raven, still circling above, watching the show, been asked his opinion, he might have nodded his head in approval of the pilot's approach to the landing. For the raven, like all bush experts, was well aware that a careful, precise approach was the foundation of a good landing.

Now that our friend the raven has kindly provided us with a brief bird's-eye view of a typical bush approach, let's take a look at what the pilot of the Super Cub was thinking, and doing, as he took his airplane into that short, rough, technically difficult dogleg strip.

Field evaluation

The first thing the pilot did, on reaching his landing area, was to fly past it at a low altitude and a slow but safe speed, giving it a careful inspection. Had this been the first time he had been in the area, the pilot would have made a number of evaluating passes as he built up, bit by bit, the reality/image in his mind, but since he had used the strip in the past he was able to determine in one pass the current:

* Shape of runway. This one, as is often the case, was doglegged, and flying past it the pilot was able to evaluate where the dogleg was in relation to the performance characteristics of his airplane. He saw that the dogleg, being at the far end, would work out okay both on landing and takeoff (i.e., his ground speed would be slow enough by the time he reached the dogleg on landing to make the turn comfortably; and, since he would be departing in the opposite direction, it would pose no problems there either).

* Condition of runway. Here he looked for changes, since gravel bars are always prone to change as the water levels that created them rise and fall, but he saw that there had been no changes since his last visit.

* Length of runway. Here, too, he looked for changes, since shifting water levels can often cut off part of a strip, rendering it unusable, but it was still the same length.

* Slope of runway. This strip had a slight downstream slope, which he noted had not changed, but it was so minor that, operationally, it didn't matter. (Actually, it was slightly in his favor, for he would be landing upstream, and taking off downstream.) There are strips, though—one is mentioned in Chapter 25—where slope is a critical element, and must be given careful consideration.

* Obstacles. There are a number of obstacles that have to be taken into consideration when evaluating a strip, both a new one, and one that has been used in the past:

Brush. Brush grows, and a strip that was once usable may no longer be usable because of new growth. Bush pilots often keep the brush cut back on strips they use regularly, and sometimes even carry a pair of heavy-duty gardening shears in the plane for this purpose.

Trees. Trees often line creeks, and they have to be taken into consideration on the approach. With this strip, an approach over the trees would not be feasible. The approach would have to be made by first flying up the narrow creekbed, then, when coming around the bend, making the landing. A tricky approach, for the chance of a successful go-around once past the final bend was marginal at best.

Terrain. There are often high banks or cliffs to contend with, and a pilot needs to take these into consideration when constructing his approach.

* Wind/temperature conditions. The temperature is easy to determine;

just glance at the thermometer taped to the wing strut. But wind direction, unless it is strong, is a more difficult matter. And, on very short strips, even a gentle breeze can create problems.

What the pilot did was to note the lack of wind-drift of the plane as he flew past the strip, as well as the lack of wind ripples on the water. This, along with the lack of movement of the leaves on the trees, was enough to tell him that the wind was calm.

Had it not been early morning—he had chosen his time of arrival carefully—there would have been wind, and that too he would have detected from the drift of the airplane, the water, and the leaves.

The approach path

The approach path is interesting from a psychological perspective, and a discussion of it could be titled, "It's all in the mind," for determining it is basically a subconscious process that is developed over time. With practice, the subconscious learns how to calculate the approach path for you, complete with the necessary three-dimensional waypoints (windows) along its route where configuration changes—airspeed, flaps, trim, and altitude—have to be made.

You must also take the following three factors into consideration:

* The aerodynamic requirements demanded by different approach routes.

* Whether ground-effect can be used (see Chapter 18).

* The missed-approach gate, its path and obstacles. These are very important with difficult mountain and one-way strips like the one used in the example.

To these must be added two questions:

1) Can the approach be performed gracefully and elegantly? This is more important than it seems, for all correct approaches have a natural flow about them when performed properly, and when this natural flow cannot be achieved with the approach path contemplated it is an indication that something is wrong somewhere along the line and additional study is needed.

2) Can you, having landed, take off again? This question, sometimes ignored with disastrous results, necessitates the consideration of a number of factors, such as terrain, wind direction, and departure paths, and should be included in every approach analysis. This is especially true of spots

down in narrow canyons or creek beds where the approach requires a complicated pattern of slips to reach the threshold gate, or those high-altitude glacier and snowfield landings on warm sunny days when the snow may be too soft and deep for takeoff on wheel-skis.

Two final thoughts to keep in mind on this subject:

* Control of airspeed is arguably the *single most important factor* involved in an approach. A couple of knots too fast or too slow at the threshold will guarantee a less than satisfactory landing. Make a point of being precise in this area.

* The best bush landings come from a rather steep final approach, terrain permitting. Trying to make a too-flat approach—dragging it in—usually results in one of the following:

1) Undershooting the landing area.

2) Overshooting the landing area.

3) A longer ground run.

The main point to remember is that the shortness of the landing, as well as its success, will be determined more by the accuracy of the approach than by anything else.

Chapter 18

Ground Effect

Between the approach and landing phases of a flight there is an interesting transition, and this occurs when the airplane enters ground effect. Normally, on regular airstrips, you don't pay much attention to this transition and just let the airplane settle through it onto the ground, considering it nothing more than part of the flare.

However, with very short strips like the example in the previous chapter, you can put this transition phase to use, and by doing so get your airplane slowed down a bit more. This often makes the difference when conditions are tight.

The same laws of aerodynamics and physics that you used during ground effect takeoffs will come into play again, only this time on the other end of your takeoff/landing spectrum. The idea is to utilize the small but important additional carrying capacity provided by ground effect to support the airplane so you can slow it to the absolute minimum prior to touchdown. In other words, to wring out the last drop of performance capability the airplane has.

During a ground effect landing, you will make your approach exactly as you would an ordinary short field landing, with the exception that you will be using an additional window. This window (or gate) is slightly—perhaps 30 to 50 feet; it varies according to conditions—*in front of the threshold,* and the airplane's altitude and airspeed on reaching it are much like they would be when entering the flare for a normal landing. In other words, to an observer who did not know your intentions, it would almost appear as if you were about to touch down a little short.

At this point you carefully increase your angle of attack and apply enough power to hold the airplane a foot or two above the surface. This will cause the airplane to slow to a speed below power-on (out of ground effect) stalling, but the airplane will remain airborne due to ground effect. It will, however, be on the very edge of a power stall, and control input must be delicate and precise. It is important to note, though, that when done correctly this utilization of pitch and power as you go into ground effect is quite smooth and subtle, and to the observer—be he passenger or raven—nothing dramatic is going on.

As you cross the threshold, you will come back on the power and the airplane will stall. There is no flare or float, and if your altitude is miscalculated, you will drop in with all the grace of a dumped tractor. If it is not miscalculated, the airplane will settle onto the ground very smoothly; with this technique you can really "paint 'em on".

Chapter 19

Landings

There are many ways to make a good landing. However, in this chapter the focus will be on STOL landings, and the first matter that needs to be dealt with is a common myth about STOL operations.

According to this myth, which has been circulating through the industry for years, STOL landings, by definition, involve a firm (i.e., hard) arrival if they are to be truly short.

I have no idea how this odd bit of folklore came about, but I assume it originated prior to the development of sophisticated bush flying techniques and airplanes like the Super Cub and C-180/185 when any landing you could walk away from was considered a good one. This is no longer true—if it ever was—and from my experience, as well as that of other bush pilots I know, it is obvious that, given reasonably decent terrain, any STOL landing can be silky-smooth. The famous "greaser" is hardly the sole province of the corporate pilot holding his jet a few inches off the runway until it gently squeaks on so as not to spill the boss's drink, or the flight instructor burning up a few thousand feet of runway to impress his student in the 152; and a hard landing is a hard landing no matter where it occurs or what the excuse is. Therefore, it is wise to avoid hard landings whenever possible, for the airplanes get enough of a beating under the best of conditions in bush work.

However, you can carry this concept too far, and it is much better to dump the airplane safely on the end of the strip if the approach has been botched slightly than to grease it on at the halfway point and go sailing off into the bushes and trees at the far end. So your priorities are:

1) To place the airplane precisely on the end of the strip at the slowest possible speed;
2) To do it as smoothly as possible.

This is perhaps asking a bit much, but the first is required for safety, the second for professionalism.

All the techniques described in this chapter, if performed properly, will provide smooth touchdowns while using the shortest possible distance permitted by the airplane's performance envelope.

Normal landings

Normal STOL landings are made using a careful approach as described in the chapter on approaches, then touching down at minimum airspeed using full flaps with the brakes partially on.

As soon as the airplane is on the ground, the flaps are retracted and maximum braking is applied. This will require full up-elevator (stick back) to keep the tail down; if the tail starts to rise because of soft terrain or other conditions, brake pressure will have to be relaxed to the point where the tail will stay down.

Strips with varying texture

Sometimes the job will require a pilot to land in fairly unsavory areas, where the "strip" is a melange of hard gravel, gullies, rocks, and sand pits. These strips require constantly varying elevator control as well as braking during the ground run. This is the type of terrain where the Super Cub shines when equipped with high-flotation tires, and is one of the reasons it has no competitors in professional bush work.

Use of flaps—high wind conditions

On a regular airport, under high wind conditions, the traditional answer is to use partial or no flaps and make a wheel landing. This is a successful technique, and is used by all pilots of tailwheel airplanes.

However, out in the bush on short, rough strips, this is rarely possible, and a slightly different technique is called for.

With this technique the pilot makes a normal approach, with a couple extra knots of airspeed added to deal with the gusts. Then, coming over the threshold with *full flaps,* he reaches down and releases the flap lever so the flaps go to 0°. This is done just at flare, and when the timing is right the airplane drops onto the ground nicely in the three-point attitude. Then, if the strip is smooth and hard and long enough, the pilot brings the airplane up to the wheel-landing position and applies his brakes.

Landing over obstacles

Light airplane specifications often specify landing distances in terms of an approach over a 50 foot obstacle, and the speed used under these conditions is the normal 1.3 Vso.

This is hardly good enough in the bush where you might have a small gravel bar up against some trees or a bluff and current wind direction and velocity will not permit you to land in one direction and take off in the other. Under these circumstances, bush pilots use one of two different techniques, depending on the airplane being operated.

Slip. This technique is used with the Super Cub and other light airplanes, and is a fairly normal technique except that in bush operations it is started low and maintained right to ground effect.

1) Using a normal bush approach, come over the obstacles and place the airplane into a steep slip. This slip is maintained until the airplane begins to enter ground effect.

2) As ground effect is entered, straighten out the slip, and, if necessary, use a slight touch of power to cushion the landing.

Partial stall. This technique works quite well with airplanes that have effective flaps such as the Cessna 180, 185, and 206.

1) The approach is made low over the obstacles using power and an airspeed of 1.15 Vso.

2) Just as the tail clears the obstacles, come back on the power and simultaneously increase the angle of attack to place the airplane into a partial stall. This causes a very rapid sink rate, and you will need to carefully balance the airplane on the edge of this partial stall as you descend like an elevator (express model).

3) Approaching ground effect apply a slight amount of forward stick to smooth the burble over the wings and relax the partial stall, while at the same time providing a slight touch of power to cushion the landing. Then, almost instantaneously come back on both the stick and power, and if done correctly the airplane will settle on the ground at minimum speed with the obstacles, because of the very steep descent, just behind it.

Some notes on this technique:

* Operations in this area of the flight envelope, when performed close to the ground, require a sensitive and subtle feel for the airplane on the pilot's part to be safe.

* Good timing is required for the power application and removal.

Both of the above techniques are pro techniques that cannot be recommended for use by pilots without the necessary experience and judgment. They are also capable of producing paint-it-on landings that will please the perfectionists among us. Safe in the hands of a pro, they can both be considered part of the bush pilot's bag of tricks.

Landing in a turn

Many bush strips will require an oblique approach where a turn has to be made just prior to reaching the threshold and the touchdown point. The usual way to do this is to turn, straighten the airplane out, and then land. This is, in fact, the approved method, but it entails a bit of jockeying around. A much more elegant way to perform this, as well as permitting landings in much tighter quarters, is to just continue the turn in one smooth sweep through to the landing. This is an easy maneuver in the Super Cub, but is more difficult in the C-180/185 because of its control response at slow speeds combined with its heavier weight.

1) Make a normal approach, adding a few knots of additional airspeed to compensate for the angle of bank.

2) While crossing the threshold in the turn, come back on the power.

3) The airplane will then touch down on one main wheel and the tailwheel.

Because this maneuver takes place close to the ground, you will need to take real care not to hit the ground with your wingtip. Like turns after liftoff, this is a pro technique not recommended for amateurs.

This type of landing is often highly enjoyed by the customers; it gives them the feeling that they are being exposed to esoteric bush flying techniques while at the same time it maintains that necessary sense of ease and safety they require and deserve. It is not unusual for them to ask the pilot to repeat the landing for the benefit of their cameras, and this is a good public relations device for customer satisfaction.

Needless to say, it is SOP to notify them prior to landing of the technique to be used so they will not be startled, which they would be if they saw you apparently prepared to land across the creek into a pile of driftwood when the strip was obviously 30° to your left.

I use several strips that require this technique. My favorite is my moose camp on the Chitina River.

Dogleg strips

Dogleg strips are easy. You just follow the curve of the strip around the bend, and that is that. But there is one caution: where the bend is in relation to the strip is important. If you will rejoin the raven for a moment, and look at the strip down in that streambed, you will note that the dogleg was at the far end. So, when the pilot reached it, the airplane was going fairly slow, even though it had about a quarter of its landing rollout to go. Then, on takeoff, the takeoff run was performed in the opposite direction, so the bend could be taken during the first part of the roll.

With a bit of experience, you will get to know just how much of a dogleg you can handle at the various points along your landing and takeoff rolls.

Soft strips

Normal soft-field techniques are used here, for they are quite adequate for most purposes.

However, under extreme conditions, the technique described at the end of chapter 3 (The Transition Times) should be used. As it is an emergency technique, it is dealt with there rather than here.

Tricycle gear

Tricycle gear is not nearly as good for bush use as conventional gear, but the Cessna 206 is such a capable workhorse that it is taken into every place that the terrain permits. It needs a smoother and somewhat longer strip than the C-180/185—a minimum of 900 feet or so, with clear approaches, if any load is to be carried. Happily, there are a fair number of gravel bars around that qualify.

When landing, I prefer to use a normal approach, though one markedly steeper than I use with the C-180/185. And I use the elevator somewhat differently than with the C-180/185. Think of it as a balancing act:

1) On touchdown, the nosewheel is first allowed to gently touch the ground (just enough to compress the nose-strut to allow ground steering to take effect).
2) Then, the weight is progressively transferred to the main wheels as the airplane slows, using up-elevator (back stick). Being too quick with up-elevator at touchdown could cause the airplane to balloon, especially if approach speed was greater than the optimum 1.1 Vso.
3) Needless to say, flaps should be retracted right at touchdown.

(It should be noted that a lot of instructors take a very strict position on retracting the flaps during the landing roll: you are not to do it under any circumstances. The reason for this is that a lot of pilots have managed to retract the gear accidently while thinking they were retracting the flaps. And I will admit that for those who will be going from 152s to light twins to commuters to the airlines, this may be a reasonable position to take. For those pilots will rarely have to operate from seriously marginal strips.

However, bush flying is a very different game, and uses mostly fixed-gear airplanes, so this quite sensible rule has to be disregarded if you are to operate safely under these conditions.)

Section II, Part II

Skis

Skis

"Think Snow" says the bumper sticker, and soon the flakes are falling and covering our bush strips and it becomes necessary to lighten our loads to compensate for the drag on takeoff. Eventually, the snow gets too deep to operate on wheels, and the time has come to switch to skis.

While I've said little in this book about float operations, with the exception of a short discussion in Section I, both forms of landing gear share many attributes. Some of these are:

* They both use natural (unprepared) surfaces for landing and takeoff.

* They both require an understanding of rapidly changing surface conditions.

* They both lack brakes, and therefore require careful advance planning for surface operations in confined areas.

* Neither is as critical of landing attitude as a conventional gear wheelplane, therefore both have a slight tendency to encourage a relaxation of discipline during the landing phase in the hands of a non-pro.

* Both are machines whose subtlety of design, and potential for sophistication of flying technique, manage to escape the uninformed.

* Both engender strong feelings of preference in pilots, and there are many who, if they had their "druthers," would only fly skis and floats.

Chapter 20

Types of Skis

There are two basic types of skis that are used in professional bush flying:

* Hydraulic wheel-skis, which can be used on both cleared runways and snow surfaces.

* Board-skis, which are restricted to snow and ice only. The basic tradeoffs are versatility, which is often a necessity, versus performance, cost, and weight. The bush pilot naturally prefers the marked performance advantages of pure skis when operating out of difficult terrain, and usually keeps both types on hand so as to better contend with changing conditions.

Hydraulic wheel-skis. When you are forced by circumstances to operate from both maintained airports and snow fields, you must put up with the many disadvantages of wheel-skis in relation to board-skis and do the best you can with what you have. These disadvantages are:

* Cost, which is in the prohibitive range and approaches the cost of floats.

* Weight, which runs approximately 70 to 100 pounds more than board-skis, depending on the size of the airplane.

* Lack of efficiency caused by the many engineering compromises required by its dual-use configuration. These compromises, under difficult snow conditions on short strips, add dangerous, if not impossible, amounts to the required takeoff run.

* Aerodynamic drag, which causes a greater loss of airspeed, fuel economy, and range than a set of floats that weigh approximately twice as much. In stark contrast to this is the situation with board-skis, which cause only about a five knot penalty compared with wheels.

For the bush pilot who must operate off rough, rocky gravel bars with his wheel-skis in the wheels-down configuration—a common situation during spring breakup and fall freezeup when operating between a snow-free low elevation and a deep-snow high elevation—there is only one model on the market that will do the job. This is the Federal "Airglide" manufactured by Genaire Limited in St. Catharines, Ontario, Canada, and distributed in the United States by FluiDyne Engineering Corporation of Minneapolis, Minnesota.

The reason for this lies in the design of the skis. With the exception of the Airglide, all wheel-skis on the market are designed for pavement operations and raise the skis parallel to the runway and only an inch or two up when in the wheels-down configuration. As a result, this restricts the user of these skis—which work quite well on snow—to paved runways or dirt strips with rocks no larger than an inch or so in diameter.

The Airglide works differently, raising the front of the skis quite high, and, while on rough gravel bars with fair-sized rocks the heels do drag a bit, this is just part of the price you pay for their utility (see illustration 27).

Maintenance is basically limited to replacing the bottoms from time to time as they get damaged from rough ice and rocks, and the occasional replacement of the bungee cords.

Board-skis. These simple and inexpensive wheel-replacement skis are basically flat boards turned up at the front (and sometimes the rear as well). They are made out of wood, aluminum, or fiberglass with a pedestal to attach them to the axle. Bungee cords (or coil springs) with safety cables are used, as with the wheel-skis, to keep them in place while providing the necessary flexibility. The bottoms can be wood, metal, or fiberglass, but the preferred bottom, for anti-sticking reasons, is a thick piece of slick plastic attached with rivets or screws. There are a number of different makes and sizes available, and most bush pilots have distinct preferences based on their type of flying, snow-conditions, and experience level.

The very high performance obtainable from board-skis (see illustrations 14 and 17) comes from their light weight and the fact that, since they replace the wheels, no compromises in design have had to be made regarding optimum size, pedestal placement—which is very important for proper action in deep snow—or flexibility of action.

With the Super Cub, board-skis increase empty weight about 35 pounds; with the C-180/185, about 70 pounds.

Tail-skis. As a general rule, precipitation is quite light in arctic and subarctic regions due to the cold, dry airmass. As a result, the snow rarely gets deep enough to require the use of a tail-ski. The exception to this rule is along the southern coastal areas where, during spring, warm moist air off the ocean overrides the colder air of the landmass and causes wet, heavy snow. Under these conditions, the snow will sometimes get so deep that the horizontal stabilizer ends up supporting the tail of the skiplane, and at this point the tail-ski will come in handy.

There are two types of tail-ski in current use:

* The wheel-penetration tail-ski, where the tailwheel is left on the skiplane and penetrates through the ski. This model is used mostly in conjunction with wheel-skis, and is distributed by FluiDyne.

* The wheel-replacement tail-ski, where the tailwheel is removed and the ski placed on the axle, just like board-skis. This ski is manufactured by Airglas Engineering of Anchorage, Alaska.

There are two valid reasons for not using a tail-ski unless absolutely necessary:

1) A tail-ski prevents you from using the tailwheel as a brake to slow the skiplane during the landing run (slide). This technique, in which you dig the tailwheel into the snow, is especially effective on bush strips that have been packed down with snowshoes or a snowmachine, and for this reason the bush pilot—especially with the C-180/185—is hesitant to use a tail-ski unless absolutely forced to.

2) The tail-ski also becomes a negative factor on very narrow strips, as its use degrades the rudder-like steering authority of the tailwheel during

that part of the ground run where there is not enough airflow over the rudder to provide positive directional control. This is particularly important on strips that have seen a lot of traffic and as a result have deep, hard-packed ruts.

The Cessna 206. Nothing has been said in this chapter about the C-206. The reason for this is that tricycle gear airplanes, for a multitude of reasons, make lousy skiplanes. Best to stick with wheels and/or floats with this machine, and use the C-180/185 or Super Cub for winter bush jobs.

Chapter 21

Ground Handling

Ski operations provide many fine opportunities for incidents of the "fender-bender" variety, and while these are naturally undesirable in themselves, they can also become quite serious—even deadly—if you happen to be in the remote bush when they occur and the skiplane is your only way home.

The techniques to prevent these problems are all quite straightforward, and in this chapter we will consider not only the normal techniques widely used by ski pilots, but also some specialized deep bush techniques that will permit you to handle strips so narrow that they are even considered marginal for wheelplanes during the summer.

The problem of sticking

Skis have an unpleasant tendency to stick to the surface of the snow or ice after the skiplane has been parked for a period of time. While this problem spans the spectrum from mild to impossible, it's useful to divide it into two classifications, based on operational considerations:

Mild. Here the skiplane is lightly stuck, and you will be able to free it by hand or power and depart without further difficulty. The degree of freeze-down will vary depending on:

* Type of skis. Some models are very prone to this problem, while others are almost immune. This factor plays a large part in the pro's ski selection.

I have found that board skis, such as the skis Atlee Dodge manufactures for the Super Cub, cause less problems with stickdown than any other type I have tried. On the other hand, I have had difficulties with the molded fiberglass type that otherwise work so well on the C-180/185.

* Type of surface. Sticky snow conditions and very cold temperatures aggravate the problem.

Mild freeze-down can occur almost instantaneously under adverse conditions but rarely progresses beyond the annoyance stage during short stop-overs of a few hours or so.

Serious. Here the skiplane is stuck so fast with compacted snow under the skis that you will be unable to depart until you jack it up and remove the accumulation. This degree of freeze-down usually takes several days to occur, and is generally the result of careless—or lazy—operating practices.

Releasing mild freeze-down

There are two techniques for this:

1) Once the skiplane is loaded, grab hold of one of the wing struts prior to getting in and rock the wings vigorously to break the skis loose from the snow. It works even better if you can get a bystander to do the job once you are in the skiplane and have the engine running, because then you avoid the risk of the skis refreezing during the starting procedure.

2) The second technique is somewhat inelegant. Get in the skiplane, start the engine, raise the tail with power, then sharply drop it. The idea is to jar the skis loose, and it often works.

Preventing serious freeze-down

There are also two methods for preventing freeze-down. These are SOP during the winter at the end of each day's flying, and ensure a trouble-free departure the following morning:

1) At home base I keep a few 2x4s—pieces of cottonwood, birch, or willow can be used—about four feet in length to block the skis off the ice and snow at my tiedown, and this resolves the problem nicely (see illustrations 15 and 31). Many pilots, in fact, carry a couple of short 2x4s in their C-180/185s, and one of these can also come in handy for turning the skiplane in restricted areas (to be covered next).

2) When flying above the tree line or over the pack ice, carry along a few 30-gallon plastic trash bags if weight or space restrictions prevent your having a couple of 2x4s along with your survival kit, for here you can not use the more southerly technique of just grabbing a branch off a spruce tree or cutting a piece of cottonwood or willow. When needed, these bags are split along the sides, opened up, and placed under the skis; sometimes easier said than done, admittedly, though necessity is the mother of invention.

If you are flying an airplane equipped with wheel-skis, and the tiedown area is ice or compacted snow, you will not need to bother with blocking the skis. All you will have to do is raise the skis so the airplane is sitting on its wheels rather than its skis (see illustrations 18 and 27).

Turning the skiplane

The normal turning radius of a skiplane is quite a bit larger than a wheelplane's and depends to a great extent on snow and ice conditions. This turning radius will not pose any problems on reasonably-sized lakes or larger airstrips, but it separates the men from the boys on narrow strips. There is a whole spectrum of techniques, ranging from the common to the esoteric, and the most useful techniques will be described here.

Normal turns. During taxiing maneuvers you can use the following two turning techniques:

1) With the tail on the ground, using rudder and power, turn the skiplane as if it were a wheelplane. This type of turn is my first choice when I have the room for it.

2) In tighter areas, and on narrow runways, a modification of the above turn is used. Here, you will give the skiplane full nose-down trim, apply power, and, raising the tail fairly high, feed in rudder and blow it around on its fulcrum. Some skis allow a tighter turn than others, and the timing of power application in relation to raising the tail and applying rudder is

important. Also important is keeping the tail clear of the snow when starting the turn, and *not allowing it to contact the snow again until all sideways motion has ceased.* The bent lower longerons on the tails of Super Cubs seen all over the place every winter are mute testimony to their pilots' lack of skill or care when performing this maneuver.

Precision turns. When the turning area is too restricted for the safe use of either of the above techniques, you can turn the skiplane on the spot where it is sitting. Part of the success of the following techniques depends upon the initial placement of the skiplane prior to the turn, for each technique requires a different amount of room and this also varies with snow and ice conditions.

* Turning it by hand. This is the easiest, and works well if the skiplane is sitting on smooth hard-packed snow or ice. Here, you push the tail around until the skiplane is facing in the direction you want to go, straighten out the tailwheel, and you are on your way.

The situation changes when the snow is deep, or badly rutted, for under these conditions if you try to turn the skiplane by hand you risk snapping the landing gear. Many pilots fail to realize that the length of the tail provides quite a powerful lever to the axles, and, were a ski to catch in a rut or be blocked by deep snow, the forces could exceed the relevant design loads.

* Hanging on a wing strut. While hardly precise, this is better than the foregoing when conditions start to get rough, for the skiplane will have forward motion and therefore the skis and landing gear will receive some protection against the turning forces. It works best with the Super Cub, though it can be used to advantage with the C-180/185.

What you do is get in the skiplane, start the engine, and then apply power as one or more of your customers holds back on one wing strut. This pivots the skiplane about its axis, and, while effective, it is hardly as precise as the following techniques, which are rarely seen outside the deep bush.

* The rope. This technique, like the one above, is fairly easy on the skis and landing gear because the skiplane is moving forward during the maneuver. It will work best with dense or compacted snow, and is suitable for all skiplanes:

1) Take one of the tiedown ropes from your emergency kit and tie a bowline knot in one end.

2) Wrap this rope around the ski on the side you wish to turn towards.

The number of loops will determine the turning radius, and it is placed in front of the pedestal but *behind* the bungee-cord and safety-cable assembly at the front of the ski. The rope should be tight enough so that under the pressure of the skiplane's forward movement it will slide back to a point just aft of the axle before the pedestal stops it. The free end of the rope is tied to the loop of the bowline knot with a slipknot so it can be removed easily after the turn. (See illustrations 36 and 37 for the general idea.)

3) Get in the skiplane, start the engine, apply nose-down trim, and give it a bit of power. Allow the skiplane to move forward until the rope slips back and catches on the pedestal. At this point, apply more power, raise the tail, and feed in full rudder. The skiplane will start its turn.

4) Continue the turn carefully, and, on reaching the point where you wish the plane to stop turning, come back on the power just enough to stop the turn while keeping the tail in the air. Once all sideways motion has ceased, let the tail down.

5) Shut down the engine, get out, and remove the rope. Your turn is now complete.

* The 2x4. This technique, a modification of the above, is used when the snow is too deep and powdery for the rope technique to work, or when it is important to make a very tight 180 degree ground turn, e.g., when using tree-lined creek bottoms barely wider than the wing-span. However, the following should be kept in mind:

1) This technique can put a strain on the ski/landing gear assembly, and is probably too severe for the Super Cub except with a very short 2x4 in exceptional circumstances.

2) While it is the best way to precisely position a C-180/185 under difficult conditions, it should be restricted to skiplanes with the proper ski axles. These are the special *eight-bolt* Cessna axles designed for serious ski operations (see illustration 37), *not* the cut-down stock axles. Failure to heed this caveat could easily result in the stripping of the axle bolts (at best) and the incapacitation of your skiplane until new bolts are installed.

Using this technique, the 2x4, which can be about a foot to a foot-and-a-half longer than the ski is wide, is placed crosswise on the snow under the front of the ski. A rope is then tied to one end of the 2x4 with a bowline or slipknot, passed over the ski between the pedestal and forward bungee-cord/safety-cable attachment fitting, and tied with a slipknot to the other end (see illustration 36). Then, the engine is started and the turn performed exactly as with "the rope" (see illustration 37).

A technique for wheel-skis in deep snow. While skiplanes equipped with wheel-skis have a noticeably smaller turning radius—due to ski length and weight-distribution patterns—than those equipped with board-skis, you will occasionally need to turn them even more sharply than possible using normal techniques. This you can easily do by taking advantage of the mechanical action of the skis when they are being retracted:

1) Get out of the skiplane and go to the side *opposite* the direction you wish to turn.

2) Take hold of the wing strut and pull down as hard as possible. This will shift the weight of the skiplane more onto the non-turning side than the turning side.

3) Go back to the cockpit, reach inside, and start pumping the skis up. The ski with the least pressure on it will retract first. When the ski on the side you wish to turn towards is in the full-up position, and the ski on the opposite side is still in the full-down position, place the wheel/ski selector in neutral. This will lock both skis hydraulically in their current position.

4) Get in the skiplane, start the engine, and make the turn as you would if using the rope or 2x4 technique. The skiplane will come around as sweet as can be, and then all you have to do is lower the raised ski and you are ready to go.

Taxiing on glare ice

Glare ice can present some interesting operational hazards for the bush pilot, and during ground operations will easily account for its share of the fender-bender accidents suffered by the unwary or careless during this time of year. The main problem here is that the surface is so slick and hard that neither the tailwheel nor the ice runners on the bottoms of the skis can get a satisfactory grip on the surface. This slickness and hardness leads to the following problems:

Starting to taxi. Admittedly, this is a somewhat unusual occurrence, and the idea that there might be a problem in getting a skiplane to *start* moving on glare ice may be greeted with raised eyebrows.

To occur, this problem requires a fairly rare combination of events. It is the by-product of the placement of ice runners on the highly esteemed Atlee Dodge board-skis used on Super Cubs. These skis have a single six-inch long ice runner in the center of each ski directly below the pedestal, and on warm spring days the black "no-slip" upper surface of these wood skis can get hot enough to cause the runners to melt into the ice. This in turn blocks one or both skis quite nicely, and, unless the skiplane is parked on a fair-sized lake and aimed in the proper direction, the standard method of laying the whip to it and bouncing the tail or wagging the rudder is not going to produce positive results. This situation is best resolved by simply moving the skiplane by hand a few feet prior to loading and departure. It takes just a slight rocking of the wings to raise each ski high enough for the ice runner to clear the hole it made, while pushing forward, to accomplish the job.

Stopping. Once the skiplane has started to move on glare ice it will demonstrate a shocking reluctance to stop, and the pilot who has not taken this into consideration might suddenly find himself in the unenviable position of being a passenger, along only for the ride, as his skiplane slides towards a row of parked airplanes.

Turning. Due to the inability of the skis or tailwheel to get a grip on the ice, turns are very difficult—and sometimes impossible—unless you blow the tail around with power. This power-turn can be tricky in confined areas unless a strong wind is present to weathercock the skiplane in the direction you wish to turn and hold it on a steady heading.

As a result, in confined areas the best technique is to plan a stop at the point where the turn is to be made, then shut the engine off (usually during the deceleration slide), get out, and turn the airplane by hand.

Taxiing crosswind. Glare ice, in strong crosswinds, makes a pilot wish it were summer and the skiplane on floats, for above a certain velocity taxiing becomes impossible, and sailing and tacking, no matter how tiresome, is better than closing up shop for the day and taking an unscheduled holiday.

Taxiing downwind. When taxiing downwind, remember that when the tailwind is stronger than the forward speed plus propwash it will cause the rudder to work in reverse, and due to the slick ice the tailwheel will not have enough purchase to prevent the tail from swinging. On a narrow runway or in confined areas you could end up in trouble.

I discovered this one sunny spring day many years ago when going into Chitina with my wife on a gas and supply run in the C-180. The runway was glare ice, with a nice little layer of flowing surface water provided by the warm 41 degree F/5 degree C temperature, and a 20 knot wind down runway 13.

I landed carefully, slid, came to a stop, and got the skiplane turned around by hand with help from my wife. Then I started to (idle) taxi back to the passenger loading area where my fuel dump was.

Everything went fine until a drift to the right developed. I applied a touch of left rudder; then, as the drift increased, I applied a bit more. Suddenly we found ourselves heading sharply to the right, toward a nasty drop off the side of the runway into the trees below, so instantaneously—by reflex, not forethought—I opened the throttle, blasting the skiplane 210° to the left and into the wind. The C-180, on that slick ice, spun like a top within its own wingspan, demonstrating extreme agility, along with a mind of its own, and came to a stop facing down the runway directly on the center line. "Interesting maneuver," my wife remarked dryly.

I decided to leave the skiplane where it sat in the middle of the little-used runway—better to quit while ahead—and haul the cans of fuel by hand from the storage tank a thousand feet away. This was probably a good idea, for the area where we were parked turned out to be so slick with its slightly downhill run that part of our job had to be accomplished on hands and knees.

Chapter 22

Takeoffs

As long as we have good snow or ice conditions a takeoff performed on skis is a real pleasure. The technique is quite simple:

1) Line up with the runway.

2) Apply flaps if desired.

3) *Smoothly* apply power (especially important on ice).

4) At the proper point bring the skiplane to a slightly tail-low attitude.

5) Allow it to fly off.

Ski operations, when snow and ice conditions are good, are far easier than wheels or floats. But when conditions get difficult, and start changing rapidly—therefore requiring a rapidly changing technique—things become different and the pilot has to truly know what he is doing.

As ski techniques so closely parallel wheel techniques we will, in this and the next chapter, assume they are the same, unless specifically noted otherwise.

Length required

This can vary from about 10% less than what would be required on wheels (that is, in ideal conditions of hard-packed snow or ice at about 36° F/2° C) to infinity in very deep snow or overflow. Some of the many factors that have to be taken into consideration are:

* Ice type (salt or fresh water).
* Snow consistency and depth.
* Surface texture (from smooth to rutted to large drifts).
* Temperature.
* Make and model of airplane.
* Type of skis.
* Condition of ski bottoms.
* Load.
* Pilot experience and ability.

All these factors will have to be correlated through experience, and it would best serve the pilot new to ski operations to take it easy until a feel is developed for terrain-ski interactions. About three seasons will give you a good basis for judgment, and at that point critical-area operations become feasible.

Attitude

For maximum short-field takeoff performance in deep snow, attitude is critically important. During the takeoff run, you will not only be developing lift on the wing, but will as well be taking the skis from a position of rest under the surface, similar to a floatplane in the displacement phase, to a planing condition similar to water-skiing. Since wing-lift is dependent on airflow, and ski planing is dependent on surface speed combined with the variables of snow texture, the adjustments of attitude must be continually adapted to the changing conditions for maximum acceleration to be obtained. This is done by feel, and the various factors operating are quite obvious to the experienced skiplane pilot.

Use of flaps

Hard surfaces. On surfaces like compacted snow or smooth ice, maximum short-field performance is obtained by leaving the flaps at 0°—or 10°

if desired in the C-180/185—until the liftoff point, then applying them as described in Chapter 16.

Soft or very rough surfaces. Very rough ice, drifts, deep snow, or overflow conditions all require individual calculations to determine which flap setting will be most effective. As a general rule, full flaps in the Super Cub, and 20°—going to the maximum permitted by load when it's time to lift into ground effect—in the C-180/185, seem to work best. These snow and ice conditions should be approached with caution by the pilot new to skis, for rough ice and drifts, if not properly handled, could cause loss of control or structural damage, and deep snow or overflow could prevent takeoff.

Turning takeoffs

The ability to make turning takeoffs becomes especially valuable during ski operations, which are performed much the same as on wheels, except that the amount of side drift will vary with snow or ice conditions.

Special problems

Glare ice. Takeoff from glare ice requires careful application of power until there is enough airflow over the vertical stabilizer and rudder to provide adequate directional control. As almost all control will be aerodynamic, many pilots prefer to maintain a three-point attitude, with the tailwheel no more than a couple of inches off the ice, during most of the run to avoid the forces introduced by raising the tail.

During strong crosswinds the skis can be operated as crosswind gear, and takeoffs are often performed with the skiplane tracking partially sideways along the ice.

Deep snow. When the snow is too deep to allow you to take off in the room available through use of the techniques previously presented, you will have to prepare the surface before you can depart. This may be done as follows:

* In snow that is not too deep to allow the skiplane to move forward, you can pack a trail down with your skis. There are two dangers to watch for here:

1) Extended ground operations at high power settings, which this type of

work requires, can easily overheat the engine. A close watch on the cylinder head temperature gauge is mandatory.

2) After making the run straight ahead, the turn back to your tracks must be made with care so as not to bury a ski in the snow (which will bring you to a stop and require digging out the skiplane). If a ski does start to drop, don't stop or come back on the power, but instead straighten out the turn a bit and bring the skiplane around at a more gentle angle. In very deep snow, or on narrow strips like the one at my home base, it is often wise to take the trouble to pack down a turning area with snowshoes, then use the 2x4 method to get back into the tracks.

* In snow that is too deep to allow the skiplane to move under its own power, get out the snowshoes and pack down a runway the hard way. Sometimes conditions are so bad that you will have to wait overnight for the snow to set up enough to allow you to depart.

* When supplying trappers, miners, etc., always instruct them to prepare a strip for you with their snowmachines. This is done by packing down a runway the desired width and length, then lining the edges with spruce boughs so you can spot it under adverse conditions (whiteout, moonlit nights, snow storms, etc.).

Drifts. These set up as hard as concrete, and can be very rough on a skiplane. The best way to take off from them is:

1) Have as light a load as possible.

2) Track parallel to the drifts, if possible.

3) Use very careful and precise attitude control, to lessen the shock loads as much as possible.

4) Get into ground-effect as soon as possible.

There is a lot of this *as possible/if possible* in drift operations, and a question that will often occur is, "Is it worth doing?"

To answer that question, you'll have to balance the importance of the mission against the wear-and-tear on the equipment (not to mention the possibility of outright damage). If you decide to go, add a surcharge to the charter fee when you deem it necessary. Drift operations will, over time, spring out the landing gear of the C-180/185, and in mild cases can require realignment of the axles. Serious cases will require replacement of the landing gear.

Finally, board-skis are more flexible in their action than wheel-skis, and therefore handle drifts better. For this reason they are preferred for drift operations and should be used whenever possible.

Overflow. Overflow on a lake is often erratic in the areas it affects. By checking things out, you can sometimes locate a dry area to operate from. If not, the best thing to do is pack down an area with snowshoes or a snowmachine so the insulating qualities of the snow are destroyed. This will often allow a sheet of ice to form that will be adequate to operate from in a few days.

Like drift operations, board-skis perform better in overflow conditions than wheel-skis, due to their better flotation and planing characteristics, and should be used whenever possible.

Chapter 23

Landings

Landings on skis are performed exactly as on wheels. The main considerations are environmental.

Length required

Like takeoff, ski landings can be quite variable, ranging from about half the normal landing distance for wheels when in deep snow to about three to four times this figure when on glare ice. On average, though, ski landings require about 20% more landing distance under conditions of hard-packed snow at temperatures of about 23° F/-5° C than normal wheel landings.

Flaps

When landing on wheels, you get your best short-field results by retracting the flaps right after touchdown, to have better braking action. On skis, since you don't have brakes, it is best to leave the flaps down for aerodynamic braking.

The "S" turn

This is a technique used to slow skiplanes on very short strips when surface conditions are satisfactory for takeoff, but somewhat marginal for landing. Basically, it calls for a series of roughly 20 to 30 degree turns made alternately to the left and right going from one side of the landing strip to the other—a snake-dance, if you will—to bleed off speed through friction while also providing a slightly longer ground run.

The "S" turn technique also comes in handy during "emergencies" when the surface turns out to be a bit slicker than you thought and the strip a bit shorter than you'd like.

A little experimentation on a large lake will help you determine just how well the "S" turn works with your skiplane—it is often quite useful with the C-180/185, much less so with the Super Cub—though it should be kept in mind that the effectiveness of the technique depends considerably on snow or ice conditions.

Special problems

How deep is the snow? Is there overflow? These are important considerations for any remote area landing where you wish to depart shortly after your arrival, for really deep snow could trap you for several days—until you're able to get a strip packed down, and it has a chance to set up—and overflow could trap you for several weeks under severe conditions.

The best way to determine the answers to these questions prior to landing is to drag your skis through the snow, then circle around for a look. This is done as follows:

1) Set up a normal landing approach with the trim and flaps set at the takeoff position.
2) Let the skiplane settle onto the lake with a slightly higher speed than normal and drag the skis through the snow. Let them get fairly deep, and add enough power to maintain your speed —much like taxiing on the "step" in a floatplane.
3) Lift off, go around, and take a look at your tracks. This should

give you a good idea of depth if ski drag hasn't already provided that answer, and water will show its presence by dark tracks.

This technique also serves to pack down a departure path for the skis, and when length allows you should land in such a way that you come to a stop in the tracks with enough distance ahead of the skiplane for takeoff. In really deep snow it helps to make several passes so the tracks are well compacted prior to landing. By doing this, you will be able to easily operate out of snow that would otherwise be far too deep for your skiplane.

Hidden overflow. Sometimes, due to a hard crust of snow or ice under the surface of the top layer(s) of snow, the above technique will not inform you of the presence of overflow. You will only discover it when the skiplane slows down enough to allow the skis to break through. When this happens there will be a sudden deceleration that requires the following immediate go-around action:

1) Full power.
2) Attitude control for best acceleration.
3) Careful flap adjustment for best lift (monitored by hand until you are free of the snow and at best angle of climb speed).

When conditions are favorable for this type of overflow, you must plan your approach and landing carefully. It is important to have a fairly long distance ahead of your planned "come-to-a-stop" point for emergency takeoff purposes, even if it requires landing at a point on the lake a half-mile or so from where your customer wants to go. As an example, there is a lake in the Wrangell Mountains where I sometimes have to land that is connected to another lake by a narrow, winding channel. When there is a potential for this type of problem, I make it a point to land in such a way that I have the mile run from one lake to the other in front of me for emergency use—and I have had to use all of it on occasion to avoid a long, wet camping trip.

Hard and rough surfaces. Here you should make a point of setting the skiplane down onto the surface as slowly and gently as possible under controlled power to limit any shock forces. On skis you do not have the shock-absorbing qualities of high-flotation tires to smooth the rough spots, so you must take special care to maintain the long-term integrity of your airframe, just as you do with floats.

Glare ice. When landing on this type of surface it often feels as though you *accelerate* once you touch down. This yields the unpleasant impression—especially with heavier skiplanes—that your ground-run might take you halfway across the state (or Territories) before you finally come to a stop. On this type of surface, all bets are off, and the really short strips are temporarily closed to your use.

As all directional control will be aerodynamic once you are on the ice, it is important to have the skiplane tracking correctly through inertia alone by the time you have slowed to the point where airflow over the rudder becomes ineffective for steering purposes.

Strong crosswinds can pose interesting problems on narrow strips, and the better part of valor under serious conditions is to pass them up until the wind dies down or changes direction.

A comment on skiplanes and short strips

My home strip is 800 feet in length, and over many years of winter operations I have always found it adequate for the C-180/185 when using board skis. I've had to use the full spectrum of techniques described above, except for the one on overflow, and I've had to keep the strip maintained. If the snow got too deep, I would pack it down using snowshoes or a snowmachine. If the drifts got too big, I would cut them down to size with a shovel. If the ruts got too deep, I would look to the sky and hope for a heavy snowfall, which, without heavy equipment, was the best I could do.

Skis make a light airplane very versatile, and, when winter comes, a whole new—and strikingly beautiful—world that wasn't available during summer opens up. So, along with many bush residents, I find myself in complete agreement with the Innuit (Eskimos): when the ground is frozen and the snow lies deep and silence settles over the land, that is the best part of the year.

Section III

Images

Caution—this section contains *advanced techniques* that could be hazardous to nonproficient or poorly-trained pilots. Refer to *Disclaimer and Warning* on page v prior to attempting any maneuvers described in this book.

Images

As bush pilots, we must accept the unpleasant fact that psychological considerations have an important bearing on safety. It is well known that most aviation accidents are caused by errors of judgment, and most of these errors are rooted in the pilot's attitude.

To improve our safety record, we must improve our attitude. Impatience with weather, get-home-itis, machismo, a lack of care or forethought—these are our deadly enemies, the foes we must defeat.

To help you understand the elements that go into a healthy attitude, I'd like to place this discussion in its natural context and take a look at some of the situations (and choices) the bush pilot encounters in his daily work. To do this, I'm going to shift gears slightly and utilize writing techniques (imagery, dialogue, foreshadowing) that are more common to fiction than nonfiction. In this way, I hope to pose problems the bush pilot faces more vividly and, by so doing, help you to understand the mindset demanded for maximum safety.

This section provided me the most pleasure in writing, for it is here that I was able to escape from the dry textbook mode into the reality of what flying and living in the North is really like, what it means to be a bush pilot, and why remote-area operations (and the associated life-style) are such an incredible joy.

Chapter 24

Survival Fit & Kit

"'Good VFR,' huh?" my wife commented dryly, gazing through the windshield of the Cessna 180 at the snowstorm that surrounded us, her voice mimicking the weather briefer in Anchorage. "'Won't have any problems,' eh? Nice to know."

"Now, now," I said mildly, "you know how weather briefings are. They have to be taken with a grain of salt. Especially in the mountains."

"More like a truckload," she responded.

And I nodded, as we continued to creep from lake to lake at half-flaps, the visibility going up and down like a yo-yo, never leaving one landing spot until I was sure we could make it to the next (and glad the airplane was on skis, it was mid-winter, and the lakes well frozen). For while we had left Anchorage in good weather, and it had remained good as far as Chitina, the Chitina Valley was blocked with unforecast snow squalls and the day was growing late. We were taking a calculated risk in trying to make it home before dark, and we still had 50 miles to go. At our current groundspeed, with no more delays, that meant another 36 minutes…

"We're not going to make it, are we?" my wife observed quietly with a glance at the clock and another at the storm outside. To either side of us the walls of the narrow valley rose, in places to an altitude of over 16,000 feet, and around us were dramatic ice fields and glaciers and stunning vistas, but none of this was evident from where we sat. We were flying in a tight little visibility circle the equivalent of ILS minimums. And while we could leave the valley from our home base in the Nizina Valley on instruments using a private SID, we couldn't go on instruments from where we were. For all practical purposes we might as well be back in the '30's with silk scarves flying in the wind…

"No," I said as we crossed the end of the lake we were approaching and the storm closed in again. I came back off the power and applied full flaps. "This squall takes us over our time limit. We'll have to RON here."

From the corner of my eye I could see her shrug. She hadn't expected any less. Then the skis were gliding smoothly through the deep snow and we parked next to a spot where there was wood for a fire and a bit of shelter from the wind. In the intense silence after the engine died the snow swirled madly and the dark came rapidly upon us. A slight gust rocked the wings gently, then subsided. There was nobody within many miles of the lake and the temperature was -2° F/-19° C. The sense of solitude was profound. We were, as one always is in the deep bush, on our own.

There are some basic requirements for flying in the north, and one of them is the survival kit. These kits contain different items depending on where the pilot is operating and the time of the year, but basically there are summer kits and winter kits. Let's look at what is necessary, or at least advisable, for flying during winter in areas where some firewood is available. The wise pilot always carefully considers the contents of his survival kit, for when the going gets tough one of the ways to contend with the often changing and notoriously difficult weather conditions is to simply set down some place and wait it out. To the bush pilot, that "some place" could as easily be a gravel bar, snow field or frozen lake as a legitimate airstrip with a convenient hotel where hot coffee and a good dinner are waiting.

So, with this in mind, let's look over the items that need to be in a winter survival kit. It goes without saying that you should get the best quality items you can afford and not skimp. When you need your kit you need it badly, and quality differences can not only provide some comfort, but also make the difference between life and death. The reason for this is because comfort has a subtle psychological consideration going for it beyond the merely sybaritic, since the ability to pitch a comfortable camp goes a long way towards encouraging a sane and rational attitude towards safe weather operating practices.

* Sleeping bags. In the interior it can get down to -70° F/-57° C at times, and, while that is an extreme, it mandates sleeping bags of the highest quality. Mummy bags are both more compact and warmer than rectangular bags, so they are the bags of choice. A good way to go is to get one rated to -30° F/-34° C as the main bag, and then get a slightly lighter one as an inner bag. This light one can also serve as a summer bag. Any of the top quality goosedown bags, such as those sold by Recreational Equipment, Inc., and North Face, will do the job well.

* Mats. A mat is needed under the bag, and a good quality foam is the type to get. Air mattresses should not be used, for they do not provide adequate insulation against the cold that will be seeping up from the ground.

* Tent. The best tents for airplane use are the type that have an external frame which permits them to be erected without stakes on rock, frozen gravel, or ice. Over the external frame is a separate rain fly, and the space between the two walls allows water vapor to escape. I currently use North Face VE-24 tents in my survival kits.

* Ax. Something is needed to cut wood, and a Hudson Bay ax, with a sheath, will do the job well. I keep a short file in my tool kit for dressing the prop when necessary, and this file also keeps my ax in good condition.

* Tool kit. All airplanes operated in the bush must have a small tool kit on board. Consider how you would feel if, just prior to takeoff, you discovered a small mechanical problem (perhaps with your preheating equipment, or with a ski bungee) that could easily be fixed in a few minutes...if only you had not left your tool kit at home base. And you are now facing a 110 mile hike out through deep snow in thirty-below-zero temperatures to get it...

* Hunting knife. A small one will do; the one I use has a 4" blade.

* Packboards or knapsacks. You will need something to store your survival kit in, and these will serve a double purpose, for if you have to walk out, they will permit you to carry your kit with you.

* Food and cooking utensils. Freeze-dried is the best choice since it is the lightest. A small nested cooking set can be purchased at almost any sporting goods store.

* Medical kit. Its need is obvious. Ask your doctor to help you put one together; remember, you might have to deal with severe pain and other serious problems.

* Flashlight, matches in a waterproof container, nylon cord, snare wire, and fishing line. Throw these in your emergency kit too.

* Gun. This is required by state law in Alaska. If you are able to place your shots well, a single-shot Thompson Contender handgun with two barrels would be a weight-saving choice. One barrel in .35 Remington will do nicely for moose and caribou, and one in .22 LR will take care of small game. Another option would be two handguns, a .44 magnum for large game and a .22 for small. However, guns are very much a matter of personal choice, and you will have to make your own decisions on this score (note: in Canada, where handguns are prohibited, your choice is limited to either a rifle or shotgun).

* Snowshoes. I prefer the trail model, for they work best for either packing down deep snow—under some conditions necessary after a heavy snowfall so one can take off—or, in an emergency, walking out. But the bearpaw style is smaller and lighter, and so is favored by many. Some of the new lightweight aluminum frame designs are finding their way into more and more kits; however, I have not yet tried them and so am unable to make any suggestions as to their suitability.

* Clothing. It MUST be a cardinal rule in winter operations NEVER to allow a passenger on board who is not dressed for the outside weather along the route. This always means parka, boots, cap and gloves. If your passenger says, "Hell, it's only a short distance and my brother-in-law's meeting me," let your competitor take him, if he's fool enough. As for the pilot, while I have heard of cases where some pilot went down wearing only light city clothes and paid dearly for it, I find this hard to believe. No Northern pilot could be that stupid. I hope.

While this completes the basic survival kit for the people on board, you still have to consider the airplane. It too needs its survival kit:

* Tiedown ropes. A complete set should always be brought along, for high winds are one of the major weather problems in the north. It would be a less than satisfactory experience to make a safe emergency landing in strong, wildly-gusting winds, only to end up having to watch your plane go cartwheeling down the lake a few minutes later because ropes were not on board.

Survival kit ropes need to be longer than those used at home base, for in the bush you will often have to use natural features like trees, bushes, driftwood, etc., or make your own tiedown anchors. During winter, a good technique is to freeze three pieces of wood (e.g, lengths of birch, cottonwood, or thin spruce about twenty-four inches long) to the ground by digging through the snow, placing them in the holes with a piece of rope tied to each, then filling the holes with snow and pouring water over them (see Chapter 27). In a sudden windstorm, this type of tiedown has held my plane secure in gusts of well over 100 knots.

* Wing and engine covers. You need these for obvious reasons; they should not be left back at home base.

* Preheating equipment. There are many ways to skin this cat. I prefer an old-fashioned blowtorch—two with the Cessna 180/185—that can use avgas as a fuel, along with three lengths of 3" stovepipe (see Chapter 12). One end of the stovepipe is placed in the bottom of the cowling, and the other end receives the torches (see illustrations 28 and 31). This method works very well down to temperatures of -50° F/-46° C, is light and fast, and the blowtorches have the added advantage of being able to help melt snow and start a greenwood fire. One final point: keep your fire extinguisher handy. While I have never needed mine, you never know…

To this basic kit (people and airplanes), always add those odds and ends which personal experience, or your spouse, suggests are useful. And, since you always have weight and bulk to take into consideration, your decisions here will often approach the Solomonic.

Which brings us to a matter of psychological import. It is necessary to consider your survival kit as just another part of the airplane's basic equipment list, like floats or wheel-skis, and treat it as such. You need it to operate in the North, and that's the way it goes. The room and weight left over is for payload, and if an extra trip must be made every once in a while because of the kit's presence, so be it. Only a fool risks leaving it behind because of customer, boss, or family demands.

The snow began to fall thickly as I put the plane away for the night, blocking the skis, putting the engine and wing covers on, and making tiedowns. And soon the only light was from the campfire where my wife was making dinner, the tent set up behind her with mats and sleeping bags already placed. Then she was beside me with a cup of coffee, and, as I took a sip, I made a final inspection of the airplane before accompanying her back to the fire. Beyond its circle of light the night was dark, and the snow made small hissing sounds as it fell into the flames…

Dinner was not freeze-dried from our survival kit, but salmon (the gift of a friend), potatoes, and vegetables, with a salad and cornbread and more hot coffee. Then it was time for bed after a long and full day, settling into our sleeping bags in the snug tent, and I was asleep almost before my head hit the rolled-up parka that served as my pillow…

…only to awake, what seemed just moments later, to find it early morning. As I opened the flap, the moonlight flowed in, casting gentle shadows, outlining the airplane waiting on the lake covered in its mantle of fresh snow. Beyond it, the mountains rose, clear and sharply etched, and above them the stars shone down with a hard brilliance. The air had the dense quality of deep cold, and, as I got dressed and put my mukluks on, it stung my hands, sending me reaching for my gloves.

Pouring a cup of hot coffee from the thermos my wife had thoughtfully filled just before going to bed, I wandered out onto the lake and over to the airplane for a look at the thermometer attached to the wing strut. In the glow of my pocket flashlight it read -44° F/-42° C.

Obviously, a high pressure system had moved in during the night, bringing with it a potential problem: the river in front of my place was still open in spots and this type of situation—very cold, dry air over "warm" water—was perfect for the formation of steam fog. Such fog, often forming very rapidly just after sunrise, could easily spread from one side of the narrow valley to the other, locking us out of our airstrip and forcing us to RON once again (although this time there was a friend's cabin nearby, which, safe from the fog due to its location, could be used if necessary). There was no question but that we had to get moving, and fast. It was going to get cold, seriously cold, and I had a feeling we were in for a spell of -60° F/-51° C or worse. A spell that could last weeks.

So, stepping out from under the wing, on my way back to the tent to start the fire, wake my wife, and begin the process of preheating the airplane's engine and getting the wing covers and snow off, my thoughts occupied with the task at hand, I moved around the tiedown rope—and, as I did, something about the cast of the moonlight on the snow attracted my attention, its ethereal beauty, the way it appeared to lie not only over but *inside* the flakes, lighting them softly from within…and it seemed almost as if…as if…

…I found I could not express my thoughts as my gaze moved over the lake in its stillness, taking it in, then climbed past the rugged mountains to the sky. And as I focused on the stars through the shimmering Northern Lights and the distorting lens of spacetime, the stars seemed to be asking the old questions, questions we have always had with us, but which now seemed as fresh, as new, as if whispered for the first time: who are we? why are we here? what is reality? what is the purpose of it all?…

Surely the deepest, most important of the mysteries.

Then, from the far side of the lake, came the howl of a wolf. Behind the camp, close, another wolf responded, the sounds flowing through the still beauty in a sense of ancient timelessness.

And somehow there seemed an answer there, more felt than understood, but an answer nonetheless.

I caught a movement out of the corner of my eye as my wife came up beside me. She said nothing—she didn't have to. We stood close together gazing at our world, its beauty and wonder and mystery, until dawn washed the stars out and served notice that a new day was beginning.

A little over two hours later we were in the air, our tracks behind us on the lake, camp clean as if we had never been there, and shortly thereafter we rounded the bend and saw our place before us. Out of the open spots in the river the steam fog was rising, spreading, licking the edges of our short strip, already covering one end, and I came back off the power, lowered the flaps, and touched down gently on the fresh-fallen snow. And, as I was tying down the airplane after draining the oil and putting the wing and engine covers on, I could not help but consider the experience and reflect that "emergency landings" and "survival conditions" were not all bad.

Not for those who are prepared both physically and psychologically.

Chapter 25

Optical Illusions: Flying's Subtle Traps

"See what I mean?" I said to my customer as we passed the rams at a fair distance from the ridge. It did not require a closer look to make the determination.

"Yeah, you were right—there's nothing in here worth—oh oh..."

In the valley below us, to one side, a large cloud of orange smoke blossomed by the side of the creek bed. It was obviously an emergency signal.

"...looks like somebody needs help."

I nodded as I came off the power and the Cessna 180 started down. Sheep season was due to begin in two days and, as always, there would be problems for some.

The airplane, a Super Cub, lay on the gravel of the creek bed, its gear torn off by a gully, propeller pretzeled, and one wing tip crumpled. In front of it some 60 feet or so began the landing strip, a narrow track between boulders, water channels, and driftwood lined with brush. Slightly wider than the width of an airplane's wheels, with just room enough for the wings to clear the brush, it was a one-way strip slightly less than 500 feet long with thick alder on the upstream end.

By the wrecked Cub one man was waving a white cloth tied to a stick; in front of a small tent a second lay in a sleeping bag with something tied around his head. He waved weakly.

"It doesn't look good," my customer said as we passed over.

"Yeah," I agreed. "We're going to have to land."

"Is there room enough? It looks pretty short for a 180."

"We're light enough. And the uphill slope will help both going in and coming out. But it will be one passenger at a time."

I made my turn, thankful the wind was calm and the early morning light would not be in my eyes, lined up with the strip, and began the approach. Passing over the Cub with full flaps, just a few feet above it, I leveled off about a foot above the terrain and raised the nose slightly, adding power and bringing the 180 to its minimum in-ground-effect speed. Then our main gear was crossing the threshold, and as it did I came off the power with brakes partially on, dumped the flaps, and we were rolling to a smooth easy stop with room to spare. Some strips, while usually only suitable for Super Cub class planes, will under certain conditions of wind and temperature be usable with the C-180/185, and this was one of them.

Then the hunter was at my door, telling us what had happened—or what he thought had happened, for it turned out he was wrong—and we were on our way to assist his partner to the C-180. The pilot had struck his head against the instrument panel (no shoulder harness), had only a bloody towel for a bandage, and was in considerable pain; it would be necessary to get him to a hospital as soon as possible.

My customer would have to wait here until my return, hoping the wind would not come up so I could no longer use the C-180, but have to return to base to get the Cub. As for my other customers, I would have to radio in and notify them there was going to be a slight delay. The problem, as so often in these undershoot cases, was the classic one of optical illusions, and in bush flying, as in instrument flying, we seem to have more than our fair share.

Some years ago, a subject in a medical experiment wore for some days an optical device that made everything he looked at appear upside down. One day, upon awakening, he discovered the world once again appeared right-side up—his brain had adapted so he could function.

When we age and go to bifocals, we often get a pair of the "invisible" type, which look like normal glasses. When first trying these, every head movement makes the objects around us seem to swim and distort as in a movie depicting an LSD trip. Yet after a few days we too, like the fellow in the experiment, wake up to discover that everything is normal once again.

So, too, with runway widths. If you are used to flying into large airports with nice wide runways, then go into the bush and begin an approach to a short, very narrow strip, it seems as though you are higher than you actually are. If you fail to recognize this illusion for what it is, you will end up flying an approach that is too low, and risk landing short. This is called *runway width illusion.*

As with most things of this nature, there is another side to the coin, for if you are used to operating into very short, narrow strips, the opposite effect occurs: when presented with a large wide strip, you will have a tendency to flare high and drop it in. This is one of the reasons behind the dislike high-time bush pilots often exhibit for commercial airports, and it just goes to show that we all do best in our home territory.

In aviation, trouble rarely comes unaccompanied, and the injured pilot had more than runway width illusion to contend with. The runway also sloped upward, which exposed him as well to *slope illusion.* With an upsloping runway, or upsloping terrain, once again you get the illusion that you are higher than you actually are. The two combined are especially deadly, and account for a number of accidents each year, with the pilot, as in this case, often having no idea what actually happened.

And once again there is the coin effect, for a downsloping runway, or downsloping terrain, reverses the illusion.

Another illusion is one caused by *featureless terrain.* You see a lot of this in winter, or when landing on a strip whose approach end is on the edge of a calm, windless lake, and this too makes it seem you are higher than you actually are.

Heavy rain on the windshield also makes it seem you are higher than you actually are.

If, VFR, you attempt to go through a thin, patchy layer of fog, you can get the illusion of suddenly pitching up.

Approaching a bush strip lit by flares or lanterns over dark terrain gives the illusion of being closer to the runway than you actually are—both in distance and altitude—causing you to tend to fly a higher approach than is suitable. On a short bush strip this could easily lead to trouble.

Flying over sparsely scattered cabins on the outskirts of a small town at night, using their lights to judge height as you approach an unlit strip, makes it seem that you are higher than you really are; once again this is the coin effect.

All these illusions create problems when landing.

It was late, almost dark, by the time I made my last trip into the little strip to pick up my patient customer. Yet, before we left, he wanted to show me something. So we walked over behind some brush where a small pile of twisted metal and fabric lay. There was a constant-speed prop, and some gear, wing and window parts from a Scout.

"Seems our friend wasn't the first to come to grief here," my customer said.

"No," I agreed. "And probably not the last." And, as we walked towards the C-180, I thought of how I sometimes made a less than graceful landing at a big airport, even after so many years in the business, and how sneaky those optical illusions are, especially in strange places or when one is tired. Yet, knowing about them, and being aware of their nature, goes a long way towards pulling their fangs. And I guess that is basically what safe flying is all about: knowing what problems are likely, then making sure they don't occur. Especially the ones that go beyond embarrassment and into damage.

Chapter 26

Swing Low (But Not *Too* Low), Sweet Chariot

"Oh, God, look where they're camped," Tommy groaned over the noise of the Super Cub's engine. "Dumb college kids!"

I smiled, having known where the camp would be, and the reaction it would elicit from Tommy. It was on a narrow ledge a few feet above the creek, a product of water erosion in the distant past. At its head the rugged glacier moraine lay, while along one side the tree-covered mountain slope climbed to jagged peaks where Dall Sheep could be seen feeding. Below it a bit, on the other side, the creek flowed over its rocky bed, and beyond the creek the mountains rose once again. Off the glacier the wind poured, gusts and eddies creating pools of wild turbulence, and things would only get worse as the day progressed.

"Coulda used the place downstream," Tommy grumbled, referring to the nice open spot with good approaches a mile and a half away, as I began the steep downwind turn, the sideways drift pronounced. A spot where tight 180 degree canyon turns in treacherous winds low to the ground were not necessary. But it was not to be. These kids had come up from California on a field trip (Alaska 101?) with their teacher, and were used to having things easy. They had money, and they expected service. Well, for the price I was charging for this little extra job, service was what they would get. Air drops were one of my specialties. Unfortunately, today's drop, because of the camp's location, would have to break most of the rules. Counterbalancing that disadvantage was the fact I had Tommy's skills to draw upon. Good help in this business is hard to come by.

Flying my downwind leg, I thought of how I had gone into the small strip at the entrance to the valley yesterday morning with the C-180 to pick them up and fly them out as scheduled. But the kids were not there, only a stick stuck in the ground and braced with a pile of rocks. On it, tied with shroud line and marked with surveyor's tape, was a plastic Ziploc bag. Inside the bag was a note. They were going to spend an extra ten days in the valley, and wanted me to airdrop additional supplies. I was also to make a phone call to California to notify one of their parents of the change in plans, so no one would send out a search party. There followed a long list of items to get, including fresh eggs (air drop?), film, and a couple of cases of beer. They would pay the charter rate for me to go into town and do their shopping. All this was fine with me; that was what I was in business for.

Turning into the wind, passing through my IAF window, I couldn't help grinning at the expression on Tommy's face when he heard about it, and saw the mountain of supplies which I had already packed for the drop. And now he was sitting in the back of the Super Cub with the baggage compartment stacked full, and the first bundle waiting on his lap, no doubt still shaking his head.

"Okay," I called to Tommy, by far the best man I had ever worked with on air drops, a genius on matters of trajectory, airspeed, and placement. "Open 'er up. We're ready to go." During World War II he had been an Air Force bombardier, and his lessons had certainly taken.

Tommy reached over and opened the door of the Super Cub, letting it fold down, then raised the window and pressed it into its catch. The wind roared in, picking up a little dust from the floor, and at that point I turned my complete attention to flying. When doing air drops, or game evaluations, the workload has to be split into two sections: the flying, and the dropping or evaluating. It is a two-man job. In this business, as the accident statistics make clear, the pilot who tries to do both does not survive long.

We descended over the stream bed, the wind rocking us, approaching the ledge from an angle. Out my window I could now see the kids waving, standing in a group, but out of the way, then we were over the ledge and I leveled the wings for a moment so Tommy could drop the first package out, surveyor's tape streaming. The turbulence was all I expected it to be, almost too much, then the package was out, rolling along the ground, and I wrapped the Cub around into a steep 180 degree turn, the wingtip just six or seven feet off the ground, the towering glacier moraine momentarily close enough to pick out tiny details. Then it was behind us as I climbed out and let the wind blow us down the canyon again.

"Well, that's one down," Tommy said in an amused tone of voice. "Only six more to go." I tossed a quick glance his way. It was as I expected. He had an ear-to-ear grin, all traces of his earlier moodiness gone. He was one of those guys that got more and more cheerful as the going got rougher and rougher, a trait that endeared him to me. A good man to have by one's side when the chips were down. His Air Force group thought so too.

So we made another pass, and another drop, then another and another. With each pass the wind got worse, the turbulence greater, Tommy more cheerful, and me more focused. Then we were done, and on our way back to my base of operations. But not to rest. The morning was just getting started, and we still had a long day ahead of us. If, as they say, there is no rest for the wicked, then we were in bad shape indeed; for, during the season, our days started early and ended late, and holidays were unthinkable.

(As a minor point of interest, I discovered later that the packaging and air drop were one-hundred percent successful: unlike the rules for this type of work, not one egg was broken.)

Okay, let's gird our loins and admit it: flying low is a dangerous practice, frowned on by all right-thinking pilots except those involved in military, agricultural, and bush operations. We have the grim FARs to remind us of this, and the even grimmer accident statistics. The military plows into the ground on a regular basis, the ag boys hit wires and wingtips, and every season hunters spin in over moose and bear.

Yet, it's gotta be done, no way around it. Low-level flight, sometimes referred to as nap-of-the-earth, or contour flying, is just too much of a necessity in our business. Since this is the case, it's best to learn how to do it right. The techniques are simple and straightforward, and, when combined with judgment and the appropriate level of flying skill and aircraft familiarity, become reasonably safe in the hands of an experienced pro.

But what bush situations require low-level flight? Well, here are a few:

* Inspection of terrain from the air prior to an off-airport landing.
* Landing and departure paths that, due to technical necessity, must hug the terrain.
* Air drops of supplies and equipment (a team operation).
* Animal population studies of mountain sheep, goats, caribou, moose, wolves, and bear (a team operation).
* Big game spotting by hunters (similar to the above, and also a team operation).
* Photographic missions, both still and moving (a team operation).
* Charting prospective roads or trails (a team operation).
* Navigation through remote mountain passes, or along river channels during periods of marginal VMC.

Note that out of the eight listed situations requiring low-altitude operations, over half of them involve a team effort. And of those, by far the most accidents are caused by only one activity: big game spotting. Therefore, the greater part of this chapter will focus on how to safely evaluate big game from the air for hunting purposes. But first we need to take a look at a few of the general requirements for learning this advanced skill.

General requirements

During your study of contour flying, there is a serious danger/temptation of which you must be cognizant. This danger/temptation (called "The Devil Made Me Do It!") comes from, and is, a product of the learning curve. For, as with any other complex skill, such as instrument flying, or playing a musical instrument with world-class subtlety, contour flying requires both careful study and regular practice. Unfortunately, contour flying is more akin to learning a serious musical instrument, such as the viola or cello, than it is to instrument flying, and as a result takes a lot longer before real proficiency is obtained. Low-time pilots, especially those afflicted with macho personalities, find this galling, and often push themselves beyond their current level of ability, hitting the evening news as well as the ground. It should be kept in mind that serious contour flying requires true in-depth mastery of the airplane, and this mastery does not come overnight.

In studying this subject and developing your skills, you will over time experience the following:

1) The development of a precise awareness of what degree of control lead will be required under various conditions of weight and ground speed to pass over or around obstacles while maintaining the desired terrain clearances.

2) The development of a precise awareness of what horizontal clearances will be required for turns through 180 and 360° under varying conditions.

3) The development of a solid awareness of the shifting aspects of depth perception under different conditions of lighting, altitude, and terrain.

4) The development of a marked ability to maintain complete concentration on the job at hand—*which is flying the airplane*—while leaving other considerations (such as, "My God, look at the size of that ram!") to your spotter.

These skills, after a period of development that usually takes a few years, along with a couple of thousand hours in the various airplanes being used, eventually combine into an ability to precisely visualize your track within your mind (exactly as is done with approaches) in advance of the proposed maneuver, and this is the key to safe contour flying. How close to the terrain you can safely and consistently operate is completely dependent upon how precisely your inward vision corresponds to reality, and the best pros, when using a Super Cub, are capable of operating safely—under calm air conditions—within about three feet vertically during straight-and-level, and thirty feet horizontally during 180-degree turns.

Basic principles

There are a few basic principles involved in contour flying, and these principles will go a long way toward keeping you safe. Let's review them briefly, along with basic control techniques, before going on to their practical application in big game evaluation.

* The pass (used both in air drops and hunting) should always be set up as a landing approach, with the target just to the side of the imaginary strip.

The go-around should be well planned, with flaps, trim, prop, and power set accordingly. You will, as always, have your key windows, just as you would for a real approach.

* The pass should always be made downhill if at all possible. This is especially important in narrow canyons, and was one of the rules that had to be broken during the airdrop episode described earlier.

* Tight downwind turns in strong winds should be avoided whenever possible. If necessary, they should be performed without ground reference. Here you are making your turn in a rapidly moving body of air, and the turn has to relate to the air mass, not some spot on the ground. During the turn, sideways drift will be quite pronounced, and trying to remain over a landmark—say, a trophy moose or bear—will just lead you into tightening the turn until you stall and go spinning in. This is a real "gotcha," and manages to get one or two pilots almost every year. This, too, was one of the rules that had to be broken during the air drop, and if you think back on the chain of events, the sideways drift was every bit as pronounced as was expected.

* Whenever possible, turns should not be made after the pass until an altitude of approximately 200 feet AGL has been reached (this joins the long list of rules that had to be broken on the air drop).

Basic control techniques

Nothing is changed here from standard operating procedures. During *steady-state* contour flying, continue to observe the following principles:

* Airspeed is controlled by elevator, especially in or approaching the area of reverse command.

* Short-term excursions from the basic flight path are made with elevator at or above 1.3 Vso as long as you remain within five knots of the target airspeed.

* Long-term adjustments, e.g., going from flat to rising or descending terrain, are made with power.

Flaps. Flap settings are dependent on chosen airspeed. Under Vfe, half-flaps should generally be used for increased stability and a more nose-down flight attitude.

Trim. Trim is used exactly the same as always, to neutralize control pressure once your desired airspeed has been locked in.

Propeller pitch. When using a constant-speed propeller during contour flying, select one of the following settings:

1) Maximum RPM. Contour flying that will—or could—require full power during the maneuver will demand full flat pitch prior to its instigation. An example of this would be when you have to go down in a deep canyon for an air drop with an immediate climb-out over obstacles after the drop.

2) High green. Normal contour flying which requires a fair amount of maneuvering will work best with the prop set at the top of the green. This allows a more refined ability to adjust engine drag or power, as well as the ability to go to a fairly high manifold pressure on demand.

3) Low green. Run-of-the-mill contour flying, as you might do under a clearly-defined 200 foot ceiling down a flat river valley over gravel bars, will not require anything but your normal cruise power settings.

Flatland turns

Prior to making any but the most shallow banks during low-level flight, e.g., around gentle river bends, or when lifting a wing over a bush or small tree, you are going to have to climb high enough to make sure you do not drag a wing tip. For most airplanes operated in the bush, a general rule of thumb is to use around 25 feet—half wing-span, plus seven additional feet for granny. It doesn't hurt, though, to go a little higher if you are not in a takeoff or landing maneuver which precludes this, for you should make a habit of allowing yourself increased margins whenever the opportunity presents itself; this is what the more experienced pros do.

Canyon turns

Turns in narrow canyons demand a high degree of precision. Because of this the fancy diving, climbing maneuver called the chandelle which the FAA recommends for this purpose is better left to them and the flight instructors. A more appropriate maneuver is the simple power-with-flaps turn, and in a C-180/185 this is performed as follows:

1) Slow the airplane to below Vfe.

2) Set the prop for maximum RPM.

3) Apply 20° of flaps while slowing to a speed ten knots above the power-off stalling speed noted in the airplane's flight manual for a 60

degree bank using this flap setting. Note that this speed varies with the airplane's current center of gravity.

4) Move over to the side of the canyon. The right side is usually preferable so the turn can be made to the left. The distance from wing tip to canyon wall under critical conditions in still air should be no less than fifteen to twenty feet.

5) Roll into a steep turn with a bank angle of 60°. Almost full takeoff power will be required to maintain altitude and airspeed.

6) When the turn is completed return to normal flight procedures.

This type of turn, once mastered, will allow you to operate safely in a very narrow canyon. The turning radius is naturally determined by airspeed and, therefore, air-stability conditions. A little practice at altitude will demonstrate the technique and its space requirements. Experience will take care of the rest.

This winds up our review of the basics, so let's move forward in time about seven weeks from the kids' air drop and take a look at hunting season. Hunting season is always a busy and interesting time of year (in many ways, much like a circus), and a good portion of the bush pilot's yearly income is earned during this period.

(1)

We came over the ridge just as the light of early morning began to flood the narrow valley, the Cessna 180 stable and comfortable in the cool smooth air, and there the two rams were, just where Ray said they would be. Then they spotted the source of the disturbance to their solitude and began to move. Momentarily, I felt a sense of empathy, then smiled as I saw the path they were picking. They were old and wise, and had our number.

"What do you think?" Ray called to me over the sound of the engine, his two German hunters in the rear busy with their binoculars. "Can we get close to them?"

I watched the rams for a moment as they moved rapidly along a narrow ledge about 500 feet below the ridge, heading into a tight V-shaped dead end: an easy place to get an airplane in, but a hard place to get it out; and far below them was tumbled moraine leading to a small gently-flowing creek.

...And that moraine—how easy to visualize a wrecked airplane lying there. As easy as visualizing how it could come to pass...

This was one of those situations that made me wonder just how much I really knew about rams (at least the very old ones, close to the end of their days), the depths of their memory and ability to think beyond instinct. It was also one of those situations where I wished we were in the Super Cub, except the Super Cub would never catch them in time, would never reach them before they made it all the way to the back of the canyon. As it was, the C-180 would make it, barely, but was there enough of a margin for safety? The walls of the "V" were close at that point, and the 180 degree turn for departure...

...Converging patterns. Calculations going on in that calm remote part of the mind where a pilot's most serious flying is accomplished. Then the answer came up, and it said we were fortunate the air was silky smooth. Another 20 or 30 minutes...

"Yes," I said. "But we'll get only one pass, so make it do."

Ray looked at me; he, too, was a pilot (his Super Cub had had an unfortunate encounter with some squirrely winds and a gully a few days ago, which was why I was flying him); then he nodded. He knew the score, had added the numbers up, knew what they meant. Perhaps that was the reason behind the grim look on his face. It was a tight spot, no question about it. Possibly one where he would not want to be flying himself.

Then we were at our IAF, and I came back on the power, flattened the prop, set in 20° of flaps and reached my 60-degree-bank stalling speed plus five knots just as we came up on the rams, passed them at about 20 feet horizontally and five feet vertically, then rolled into the turn with almost full power.

The sloping walls of the canyon rotated around us, looking almost as though we were about to land on them, brush them with our wheels, then we were clear and headed down the valley, leaving the rams behind.

Ray turned in his seat and asked his hunters, "OK?"

Their answer was lost in the sound of the engine, but it did not matter. Even the glimpse we had caught before entering the final portion of the approach had been enough. Needless to say, their binoculars had remained unused—as both Ray and I had known all along would be the case. They had got their close look. A look perhaps somewhat closer than anticipated.

And Ray flashed me a subtle smile. He knew how tight the turn had been, how close to the walls we had had to go to accomplish it. We were fortunate our hunters had found something to distract their attention. That V-shaped area of the canyon was not a place to go if inexperienced…either as a pilot or passenger. It could be hard on one's dreams. At best.

Our destination now was a gravel bar about 30 miles away on the wrong side of the river, one suitable for a heavily-loaded C-180. I would leave them there and return to base for the Super Cub. With it I could get them within "reasonable hiking distance." And I would wish them luck—they would need it. Those two rams, old, wise in the ways of hunters, intelligent, were a worthy challenge.

(2)

Late afternoon, two weeks later. The air was smooth with a high overcast, and ahead of us was Chickaloon pass, closed on the Anchorage side due to weather. The C-180 purred along smoothly and we were starting our climb to go on-top, join the GKN-BGQ airway, then make an instrument approach into Anchorage, when one of the Germans pointed outside the plane at something he wanted to take a look at. His eyes were as good as his hunting abilities, as good as the ram he had gotten, and far below us a small brown spot moved across the muskeg, paused, then began digging at something.

So we lazily circled down in the fall colors, came to our IAF behind the bear, a fine Toklat with unusual markings, and approach began to turn into pass before he heard us, realized he was not alone, stood, and swatted at the large mosquito (Alaska size) that had suddenly shown up. Their ability to move fast, the snake-like sinuous flow of their movements, never ceases to amaze me. Then the pass was over and we were at our go-around point, starting our climb, the rule here, as always, being one pass only, for it is not good to disturb game without real reason.

"Once more," one of the Germans said. "Please." He made a slight movement with his hand. In it was a movie camera.

A reasonable request. So I continued on up to 200 feet, made the turn, passed through our new IAF, made the pass as the bear once again stood up and swatted at us, a fine animal, then we were past and had reached the go-around point. And as the power was gently applied, Ray turned around and asked, "Do you want to hunt it? We can land back at the river."

"No," said the German who had not taken the pictures. "Not this time. Maybe next year."

"Well," Ray said, "you both got good rams. One will go high in the book. I guess that is enough for one season."

The German with the camera smiled. "We will not be entering them," he said mildly. "It is the competition with oneself that counts, not others. That is all that matters."

None of us had anything to say to that.

It was the pro attitude. The best hunters had it, as did the best pilots.

Chapter 27

A Slightly More Than Gentle Breeze...

"Jesus, I can't see a damn thing," Belphi said, tightly gripping the wheel of the Ford 4x4. The wind, gusts hitting over a hundred knots, rocked the truck, and the blowing snow reduced visibility to about fifteen feet.

"A little to your right," I said as a dark shape loomed faintly off to the side, too obscure to make out. "I think we're coming up on the runway."

We worked our way closer to the shape, and something about it, though strangely distorted, struck a note, then snapped into focus. It was a neighbor's PA-16, one we knew well, on its back. Obviously, the wind had broken it from its tiedown and blown it several hundred feet to the runway and its present position.

"You called that one right," Belphi said in disgust, looking at the airplane, or rather, what was left of it. "Let's get over and see how the Cub's doing."

We were living temporarily in Wasilla where there was electricity to run a word processor so I could do some writing, and, for the first time since I came to Alaska, I was forced to base my airplane at a public airport. I found it a less than edifying experience. Especially at times like this.

As we crept around the PA-16, a gust moved it several feet further along the ground, in the direction of the fuel pumps and highway. Then we reached the row of tiedowns and began to creep slowly along it, our Super Cub still not in sight. I could feel the tension building, a form of tension I was unused to, then our airplane finally loomed out of the blowing snow. It was still in place, riding the gusts nicely as it had been doing the previous evening, and even the wing covers, which now were serving as control locks, were still attached. I had not expected otherwise, of course, but then, the way my airplane had been tied down was not the source of my concern.

"Let's see how accurate your prediction really was," Belphi said, and I nodded. I too was curious. Wasilla was one of those airports where each person installs his own tiedowns, and one day just before winter hit I went down to the airport after my morning's writing stint and discovered I had a new neighbor. It was the PA-16, and like many airplanes you encounter on public airports, it was carelessly tied down. The ropes were old, rotten and loose, and almost thin enough to be considered string. I took one look, and it didn't take a seeress with a crystal ball to read the future. Later in the year a strong winter high would build in the interior, with a deep low off the coast, and through Chickaloon Pass the wind would roar, sweep across the airstrip from the northeast, grab the PA-16, tear it loose from its tiedowns, then blow it into and over my airplane, destroying it.

So I made a point of looking the PA-16's owner up, and having a neighborly little chat with him. But his attitude was that *his* way of tying an airplane down was perfectly adequate. "Never had any problems before," he said. So I shrugged, the world being full of people like him, and moved my Super Cub down the row far enough that it was out of his way.

But this wasn't the end of it, for as soon as I moved my Super Cub, two new airplanes came onto the field. The first, an Ercoupe, parked in my old spot. I warned the fellow who owned it about what might happen, but he was unconcerned, so I said nothing further; being a good neighbor has its limits. Then, a few days later, I found a Stinson parked immediately next to my Cub. And it, too, was tied down in such a way that the first real storm would tear it loose.

A little investigation revealed its owner was a guy working on the pipeline, and that he was on the Slope. But one day as I went to check my airplane, I found a young woman at the Stinson and fell into conversation with her. Turned out it was her husband's plane, and when I explained the problem with the tiedowns, she got in her car, drove over to the hardware store, and bought some ½" nylon rope, like I used. Good woman; figured her husband, who was new to flying, would want her to take care of the problem in his absence, which turned out to be the case.

"Well, Ron's plane is okay," Belphi said as we passed the Stinson, then came up to the Ercoupe and parked. "Gonna set up a tent at the next Anchorage Street Fair? Tell fortunes?"

I said nothing, just sat there and looked at the Ercoupe in silence. Like the PA-16, there wasn't much left of it, and the tragedy was that there was no excuse for this. It takes so little to tie an airplane correctly, and the reward is so great.

"They never learn, do they," Belphi added thoughtfully. "Except for those like Ron and his wife. You gonna put a chapter in your book about this, for those who could use the information?"

"I hadn't planned on it," I said. "But you're right—it is a problem, and there certainly are a lot of pilots who don't seem to know how an airplane should be secured against high winds."

"Well, mention me in it," she said, grinning.

"Okay," I said, smiling. And we returned to the Super Cub, so I could climb out of the truck into the -85° F/-65° C wind-chill factor and re-tension the ropes, something that often needs to be done daily during the more serious storms.

There is a certain security that comes with living in the remote bush (call it one of the "perks"), and this security extends to our airplanes, even if they are parked on a remote wilderness gravel bar. True, nobody is going to come and hold your hand if your airplane gets destroyed by a powerful wind storm, but then, if that event does happen, at least you do not have to search very far to discover where the fault lies. This being so, let's take a systematic look at how permanent tiedowns should be installed at a base camp.

Location

It is nice to have some form of shelter to use as a wind break, such as a high bank or a line of trees. However, because of the shallowness of the northern spruce root system in the areas I operate from, I've always felt better having my airplanes out in the open where they can't be reached by falling trees. It's also nice to have the airplanes out in the open because of bear, for denying a bear shelter for concealment means less chance of a problem that cannot be satisfactorily resolved. (My first wife, Dolly, had a little disagreement with a grizzly one morning when walking down to the tiedowns; the bear unwisely chose to use force to resolve their dispute and ended up on the losing end—see illustrations 38 and 39 for the participants in this not uncommon mini-drama of bush life.)

Direction to face

I prefer to face my airplanes into the prevailing wind, whenever possible.

Tiedown anchors

Sometimes you will see airplanes tied to a couple of concrete blocks lying on the ground, and this surely has to be a joke. These blocks weigh about 50 pounds each, and attaching 50 or 100 pounds of weight to each wing is hardly going to do much good in a serious wind storm. (And yes, I have seen a number of airplanes rolled into a ball during wind storms that were carelessly and ignorantly tied to concrete blocks. One was a guide's Super Cub at Wasilla one breezy day in the middle of moose season—and during that minor storm the wind never got above 60 knots.)

Here is how I installed the first permanent tiedown at my home base in the Wrangell Mountains:

1) I went to an abandoned mine and selected three rusted four-cylinder engine blocks from the operation's dump. These I took to Spruce Point in the Super Cub. Through one cylinder of each block I passed a galvanized steel chain with an inside link diameter of five-eighths of an inch. The end link at the block was secured to the appropriate link on the main length of the chain with a sturdy bolt.

2) I dug three holes about three feet deep. In each hole I placed a block, with the chain coming out the hole. I then filled the holes with rocks and dirt, and compacted it well.

3) The chains were long enough to reach two feet above the surface. The reason for this length was to protect from snow and ice abrasion the nylon ropes that went from the end of each chain to the airplane's tiedown rings (see illustration 15).

A few years later, when I was able to afford a C-180 and additional tiedowns became necessary, I went through this process again, although this time, since there were no more scrap engine blocks available, I had to resort to old pumps and similar bulky, weighty junk for my anchors.

Rope (also known as line)

I use high-quality ½" twisted three-strand nylon mountain climbing rope. When the rope begins to show signs of fraying, or is over two years old, I replace it.

Knots and tension

Knots and tension will be treated together, for together they provide the strength and resiliency necessary to defend an airplane against high winds.

The bowline. Starting at the bottom, fasten your rope to the tiedown ring or chain with a simple bowline knot (see illustration 40). The bowline is the best (I'd go almost so far as to say only) knot to use for this purpose. Easy to tie, and easy to untie, the tighter the rope is pulled, the better it holds.

Tension. Once the bottom of the rope has been tied to its chain or ring, put the upper end of the rope through the wing's tiedown ring and apply a slight amount of tension. What you want the airplane to do in a wind storm is to give slightly in the gusts, but no more. It is a matter of minimizing the shock load the airplane receives from the gusts, and the combination of both the ½" nylon rope's tension, and ability to stretch slightly under strain, provides considerable protection.

A good way to gain an understanding of this process is to consider what happens to airplanes that have been tied down with their ropes slack. Go to any airport and there they are, just waiting for the next serious storm. Then afterward, as their owners survey the scene of destruction (see illustrations 41 and 42), they just shake their heads at the unfairness of fate. "Stuff happens," they say, shrugging, and continue on their way. Business as usual.

Well, "stuff" doesn't just happen. It is caused. Let's take a look at what is going on here.

On a nice day, everything is fine. The slack in the ropes allows the airplanes to rock gently in the breeze, and everything is right with the world. But then, one day or night, a serious storm arrives on the scene, and things change. As the wind picks up, the rocking becomes more pronounced, and the ropes begin to stop the wing's upward movement. Things are still okay, though, for the wind has not yet developed enough force to over-stress anything. But then the gusts get stronger, and the situation becomes critical. For when they start to hit hard (say, 75 to 100 knots), the wing swings up, then slams to an abrupt halt as the rope snaps tight. The force caused by the moving mass of the wing, combined with the force of the gust, is then focused on the small area of the wing's tiedown ring, and sooner or later, under repeated assault, something is going to give. Sometimes it is the ring, other times the rope. But whatever it is, the airplane maintenance shops are gonna smile and the insurance companies cry.

It is easy to see why chains (used by many pilots, and FBOs) are worse in this situation, for while good ½" nylon rope has a certain amount of give to lessen the shock, a chain has none.

The Hurricane Hitch

Now that you have the bottom end of your rope attached to your anchor, and the top end has been reeved through the wing's tiedown ring with a slight amount of tension applied, the rope will need to be secured. The best system of knots to use here is called the Hurricane Hitch. What you do is take two turns around the standing part of the line a couple of inches below the ring, jamming the second turn under the first to hold it in place. The tension on the rope will be held by the two turns. Make a second jammed pair of turns a few inches below the first, and then a third a few inches below that. (Illustrations 43 and 44 show clearly how this should be done.) Then the end of the rope is half-hitched just below the third jammed pair, and that's it.

Ring paranoia

Sometimes with Super Cubs and C-180/185s ring paranoia strikes, and the mistrustful soul will tie the rope around the strut rather than use the factory's tiedown rings. The correct way to do this is shown in illustration 44.

Emergency tiedowns

Any time of the year you might have to land someplace where tiedowns are not available and you will have to make your own. The worst place for this to occur is at some of those big commercial airports in Canada where provision for general aviation has deliberately been ignored. This is a tough nut to crack, because for some obscure reason they don't want you digging holes in their ramp, or during winter freezing pieces of wood to it. Fortunately, airports on the main inland route to Alaska like Prince George, Fort St. John, Fort Nelson, Watson Lake, and Whitehorse provide tiedowns, but at airports like Vancouver, Yellowknife, or Inuvik, you can forget about these little niceties.

There is no question that the bush is more friendly.

Winter. This is easy. Just dig three holes in the snow, and lay a twenty-four inch length of whatever wood is available in each one, with a rope attached. Then fill the holes with snow, compact it firmly, and pour water on it. (If an open creek isn't handy, melt some snow using your preheating equipment, or a campfire.) Shortly thereafter the snow will freeze to solid ice, and your tiedowns will be ready for use.

This system has worked very well for me over the years, and several times has easily withstood extended winds gusting over 100 knots.

Needless to say, every time this technique is used, your ropes will get a little shorter, for you have no choice but to cut them free just above the surface prior to leaving.

Summer. In summer, when the winds are not so fierce, you do not have to take the extreme precautions that you would in winter, but then, putting in a good temporary tiedown is also more difficult. I often seek out clumps of willow, birch, or cottonwood to use as tiedown anchors (see illustration 12), and for temporary use this works well. Driftwood logs, if they are heavy enough, or partially buried in sand and gravel, also work well. Otherwise, holes will have to be dug, and chunks of wood buried to serve as anchors.

Whatever the situation, you are going to want to tie your airplane down if wind is on the agenda.

"I'm glad we don't keep our 206 over there," Elaine said, gesturing out her side window as we taxied past a row of airplanes on the Tucson ramp. I was many years, another marriage, and 2,300 miles away from the incident reported earlier, and we were making like snowbirds and spending the winter in Arizona.

"Yeah," I said, casually glancing at the loose chains. "One of these days a good thunderstorm is gonna pass through, and that line will turn into a junkyard."

"Well, I'm glad you're renting a hangar, even though it costs an arm and a leg," she said. "Though at least one of those planes seems to be tied down okay."

"The yellow Cessna 140? With the ropes on it?"

"Yeah."

"That's Barbara's plane."

"Oh, no wonder," Elaine said, thinking of the beautiful, hard-nosed, and highly competent DE with whom I took my six-month instrument check rides when we both happened to be in town at the same time, I down from the arctic, and she from her airline pilot's job in Texas. "But doesn't she worry about it?"

"Sure, but what can she do? She can't afford a hangar right now, so she just has to tough it out."

Elaine sadly shook her head. Like me, she didn't need a crystal ball to read the future, and only luck was going to protect Barbara's airplane from the incompetence and carelessness of her neighbors. Unfortunately, in aviation luck can't be counted on. And in this instance, that proved to be the case.

Less than two years later, during the monsoon season, a serious thunderstorm passed over the airport and swept down the line of airplanes. Some of the results of its passing are shown in illustrations 41 and 42. As illustration 41 shows, Barbara's Cessna 140, having been properly tied down, didn't budge. But the same can't be said for the Cessna on its back, which had been loosely secured with chains; one of them can be seen hanging from its tail. Illustration 42 shows the other airplane involved in this sordid little triangle.

Bush flying, like all forms of flying, is a matter of details. And no detail is too small not to be important. There is an old saying that a flight is not over until the airplane is parked and properly tied down. Nowhere is this more true than in the bush, where help is unavailable. This is the reason the pro takes such care of seemingly minor details, and, in these details, survives.

Appendices

Appendix A

Glossary

ADF

Automatic Direction Finder. An instrument that, when tuned to a NDB, will point at the NDB with its needle.

Breakup

The time, in spring, when the ice starts to go out because of warm temperatures.

CAVU

Ceiling And Visibility Unlimited.

Chinook

A warm wind associated with a warm front during winter; said to mean "snow eater" in the Indian language for its ability to melt snow.

DH

The point on an ILS instrument approach where the pilot must have the minimum ceiling and visibility mandated by the instrument approach plate to continue the approach to a landing. At that point, if the runway environment is not in sight, further descent is not authorized, and a missed approach must be made.

FARs
Federal Aviation Regulations.

FBO
Fixed Base Operator. A facility on an airport where pilots may purchase fuel, have maintenance done, etc.

Freezeup
The time, in fall or early winter, when the ground and lakes become frozen because of cold temperatures.

Go-around
A missed approach to a landing where the pilot elects to discontinue the approach and return to normal flight.

IAF
Initial Approach Fix. This is the point where, when flying instruments, the beginning of the approach segment commences. It is also used in this book as the point where an approach to an off-airport landing—or low pass, when contour flying—is begun.

IAS
Indicated Airspeed.

IFR
Instrument Flight Rules.

ILS minimums
The minimum visibility you must have at Decision Height (DH) to continue an ILS instrument approach to landing. Usually 200 feet vertically, and one-half mile horizontally.

IMC
Instrument Meteorological Conditions.

Line
Rope. Specifically, a piece of rope that has been cut from a coil for a particular use, such as tying an airplane down.

LOM
Locator Outer Marker. An outer marker installed in conjunction with a NDB.

MEA

Minimum Enroute Altitude. The minimum altitude on an airway a pilot can fly and still have standard terrain/obstruction clearance.

Mukluks

Insulated boots, often made of canvas with felt inner liners and a rubber sole, that come almost to the knee. These are very warm in cold weather, and are the footwear of choice in the arctic.

NDB

A low frequency radio beacon. It will cause the needle of an ADF or RMI to point to it.

NTSB

National Transportation Safety Board.

NWT

Canada's Northwest Territories, an area of stunning wilderness beauty.

OAT

Outside Air Temperature.

OM

Outer Marker. A marker beacon at or near the glide slope intercept altitude of an ILS approach.

On-top

Flight above a cloud deck, often requiring an instrument approach at one's destination.

"Out"

A pilot's emergency plan if things do not go right.

Reeve

To pass a rope (line) through a tiedown ring or other attachment fitting.

RMI

Radio Magnetic Indicator. A sophisticated instrument that has two needles backed by a slaved compass card. One needle points to NDBs; the other points to VORs, as well as Loran, GPS, and RNAV waypoints. The RMI is much loved by highly experienced pilots, though it is usually hated (and held in scorn) by low-time pilots because of the (erroneously) perceived difficulty in learning to use it.

RON
Remain Over Night.

SID
Standard Instrument Departure.

SOP
Standard Operating Procedure.

Toklat
A form of grizzly bear.

TRW
Thunderstorm with rain showers.

Tundra tires
Low-pressure, high-flotation tires, designed for landing on soft or rough terrain.

Vfe
Maximum flap-extended speed.

VMC
Visual Meteorological Conditions.

Vso
Power off stall speed in the landing configuration.

1.1 Vso
Power off stall speed in the landing configuration, plus 10%.

1.3 Vso
Power off stall speed in the landing configuration, plus 30%.

Appendix B

Sources

For additional information on subjects referred to in this book, contact organizations at the following addresses:

Airglas Skis for Super Cubs and C-180/185
Airglas Engineering Co., Inc.
W. K. Landes, President
Box 190107
Anchorage, Alaska 99519
(907) 344-1450

Cessna 206 Deicing Equipment
Kohlman Aviation
319 Perry Street
Lawrence, Kansas 66044

Cessna Pilot's Association
John Frank
P. O. Box 5817
Santa Maria, CA 93445
(805) 922-2580

Flightmaster Navigation Computer
Flightmaster, Inc.
Lake Technology Park
P. O. Box 128
McHenry, MD 21541
(800) 462-6669 (301) 387-7997

Long-Range Fuel Tanks
Flint Aero
8665 Mission Gorge Rd, Bldg. D-1
Santee, CA 92071
(619) 448-1551

Long-Range Fuel Tanks
Sierra Industries, Inc.
P. O. Box 5184
Uvalde, Texas 78802
(512) 278-4381

Super Cub Mods & Skis
F. Atlee Dodge Aircraft Services, Inc.
P. O. Box 190409
Anchorage, Alaska 99519
(907) 344-1755

Super Cub Mods & Tundra Tires
Cub Crafters, Inc.
P. O. Box 9823
Yakima, WA 98909
(509) 248-9491

Super Cub Pilot's Association
Jim Richmond
P. O. Box 9823
Yakima, WA 98909

Tanis Preheat System
Tanis Aircraft Services, Inc.
P. O. Box 117
Glenwood, Minnesota 56334
(800) 443-2136

Wing and Engine Covers
Alaska Wing Covers
Linda Logan
18521 Whirlaway Road
Eagle River, AK 99577
(907) 561-5412

Wheel-skis
FluiDyne Engineering Corp.
5900 Olson Memorial Highway
Minneapolis, Minnesota 55422
(612) 544-2721

Appendix C

Index

B

C

D

E

M

N

O

P

Appendix D

Photographs

1. Author's Super Cub on Nizina River, across from Spruce Point, circa 1968.

2. Spring sunbathing, Alaska style, on the Knik River, circa 1981. Author's Super Cub and Belphi.

3. Author's Super Cub on Tana River, circa 1968.

4. Author's Super Cub on Tana River, circa 1968.

5. Author's Super Cub on Knik River, circa 1978.

6. Author's Super Cub at Fox Island, circa 1982.

7. Author's Super Cub flying over landing area on the Knik River, circa 1978. Pilot F. E. Potts. *Photo by Martin Cramer.*

8. Author's Super Cub on FEPCO Aviation's home "airport" at Spruce Point during the early 1960s. This gravel bar was subject to regular flooding (note water mark and developing gully behind airplane), and a few years later, after it had been improved with a "cat," was washed out (see photo 9).

9. Belphi standing in the middle of the FEPCO Aviation runway at Spruce Point during the flood that finally washed it out in 1971.

10. Dolly with Author's Super Cub at Hidden Creek sheep camp, circa 1969.

11. New FEPCO Aviation airstrip, one-half mile from Spruce Point cabins, c 1971.

12. Author's Super Cub, during a period of flooding of the main FEPCO Aviation airstrip, tied to two cottonwood trees in a field of fireweed, rocks, and driftwood a half-mile from the cabins, circa 1969. Note ropes faintly visible in photo, and marked with arrows.

13. *Sheep hunters breaking camp on a 500' mountain saddle strip at an elevation of 5,000 MSL, c 1968. Wing of Author's Super Cub shows in the foreground.*

14. *Author's Super Cub equipped with Dodge Board Skis on Knik Glacier Lake, circa 1979. Shortly after this picture was taken, a pilot put his C-185 through the ice in the same spot. Ice conditions can change rapidly, and must be continuously monitored.*

15. Author's Super Cub, on old FEPCO Aviation airstrip (since washed out), circa 1968. Note wind-polished snow, snowshoes tied to wing strut, and skis blocked against freeze-down. The arrows point to the chains that go from the tiedown anchors to the ropes.

16. Author's Cessna 180 on new FEPCO Aviation airstrip at Spruce Point during the summer, circa 1971. Note that airstrip is on two levels, each about 400 feet long.

17. Author's Cessna 180 on new FEPCO Aviation airstrip at Spruce Point during the winter, 1971/72. Note the Landes Airglas board skis with eight-bolt ski axles. Also note the spruce boughs marking strip boundaries behind the airplane (there are large rocks, logs, and chunks of driftwood on both sides of the strip.

18. Author's Super Cub on wheel-skis, parked in front of the new FEPCO office/ living quarters at Spruce Point, circa 1969.

19. Typical overcast spring day during breakup in Alaska. Photo taken from Author's Turbo Cessna 206, 1988. Note "rotten ice" on the lakes, especially those in the foreground.

20. Flying "on-top" through Chickaloon Pass into Anchorage in a Turbo Cessna 206. When the photo was taken, the pass had been closed for three days in snow and fog, with VFR pilots lined up waiting at Gulkana, Eureka, and Palmer for it to lift. We went on top, intending to make an instrument approach into Anchorage, but found the weather clearing in the Palmer area (though not in the pass) and let down through a large broken area over the Knik Glacier so we could take more photos. *Photo by Elaine Potts.*

21. Photo of one of the Author's cabins at Spruce Point, taken by moonlight February, 1970.

22. Preheating the Author's Super Cub at Spruce Point, circa 1968. Note the "Sun-Dog" caused by a mild ice fog condition. The temperature that morning was -40° F/-40° C.

23. *Ernie Snowden and Dolly standing by the Author's Super Cub in the Wrangell Mountains, Alaska circa 1968. Note safety-cable installation and tire inflation.*

24. *Author's Turbo Cessna 206 at May Creek, Alaska 1988. The log cabin is the local mail drop for residents of the area.* *Photo by Elaine Potts.*

25. Elaine standing next to the Author's Turbo Cessna 206, 1987.

26. Instrument panel of the Author's Turbo Cessna 206, taken in the Tucson Avionics' Hangar, Tucson, Arizona, circa 1990.

27. Wing covers on Author's Super Cub equipped with wheel-skis at Spruce Point, April 1969. Note snowshoes tied to wing strut.

28. Author's Super Cub being preheated at Wasilla Airport, winter 1978, using the blowtorch method. Note that the heat-source is on one side, and that the far-side air inlet is uncovered so the flow of heated air can remove the cold.

29. Engine cover on Author's Turbo Cessna 206 in unheated hangar. This airplane has a Tanis Super System electric engine heater that is good to -65° F/-54° C.

30. Tanis electric engine heater installation on the Author's B36TC Bonanza. Note the fitting on the baffling, and the electric extension cord plugged into it. With airplanes whose cowls open through the use of latches, like the Super Cub and the Bonanza, this is the best way to do the job.

31. Author's Cessna 180 on Landes Airglas board skis being preheated in the bush, winter 1971, using the double-blow-torch method. Note how the skis are blocked to prevent freeze-down; and the eight-bolt ski axles.

32. The mountain pass from the White River into the Author's base of operations at Spruce Point, circa 1972. *Photo taken from the Cessna 180 by Belphi Potts.*

33. Author's Turbo Cessna 206 entering Chickaloon Pass from the Eureka side, 1988.

34. Chitina River where it meets the Nizina River, circa 1971. Taken from the Cessna 180.

35. Author contour flying the Turbo Cessna 206 up a creek in Alaska during spring breakup, 1988. *Photo by Elaine Potts.*

36. Initial position of the 2x4 and rope on a bush short-radius ski turn. Note the position of the rope in relation to the ski cables, along with the rope's tension.

37. Final position of the 2x4 and rope after turn. The distance of the 2x4 behind the pedestal is important, and the rope should initially be adjusted so that, when the airplane is moved forward, the 2x4 ends up in the proper position. Note also the special eight-bolt ski axle.

38. Dolly in front of FEPCO Aviation's first home, circa 1968.

39. Dolly with the bear that attacked her, circa 1969.

40. Bottom end of a tiedown rope attached to a public airport's anchor. Knot is the common bowline. Note that the airport chains were not used.

41. Puzzle time: find the three airplanes in this photo. The Cessna 140 on the left belongs to a friend, and was correctly tied down. Also note the broken tiedown chain hanging from the tail of the Cessna on the far right.

Photo courtesy of Dick Rose, Hotton Aviation, Tucson, Arizona.

42. Another view of the same event shown in illustration 41.

Photo courtesy of Dick Rose, Hotton Aviation, Tucson, Arizona.

43. "Hurricane hitch."

44. Proper way for those with "ring paranoia" to tie their tiedown rope to the strut of a Cessna 185, instead of to the tiedown ring. Note how one loop of the rope goes through the tiedown ring to prevent the rope from working down the strut during a strong storm.

45.Author at Hidden Creek, Alaska, 1968.

Photo by Dolly Potts.

46. Author at May Creek, Alaska, 1988.

Photo by Elaine Potts.